T0262704

Protein Interactions: Selected Topics

Protein Interactions: Selected Topics

Edited by **Anton Torres**

New York

Published by Callisto Reference,
106 Park Avenue, Suite 200,
New York, NY 10016, USA
www.callistoreference.com

Protein Interactions: Selected Topics
Edited by Anton Torres

International Standard Book Number: 978-1-63239-521-4 (Hardback)

Contents

Permissions

List of Contributors

Preface

Protein interactions are being studied by researchers across the globe. This book presents the interaction between proteins and other biomolecules, which is crucial to all features of biological procedures such as cell growth and differentiation. Hence, examination and modulation of protein interactions are of vital importance, as it not only discloses the mechanism governing cellular movement, but also leads to possible agents for the treatment of a broad range of disorders. The purpose of this book is to emphasize on some of the most recent advances in the study of protein interactions, along with modulation of protein interactions, improvements in systematic methods, etc. It presents diverse approaches for the study of vital protein interactions. This book demonstrates the significance and the opportunities for further exploration of protein interactions.

Various studies have approached the subject by analyzing it with a single perspective, but the present book provides diverse methodologies and techniques to address this field. This book contains theories and applications needed for understanding the subject from different perspectives. The aim is to keep the readers informed about the progresses in the field; therefore, the contributions were carefully examined to compile novel researches by specialists from across the globe.

Indeed, the job of the editor is the most crucial and challenging in compiling all chapters into a single book. In the end, I would extend my sincere thanks to the chapter authors for their profound work. I am also thankful for the support provided by my family and colleagues during the compilation of this book.

Editor

Studying Protein Interactions

Live In-Cell Visualization of Proteins Using Super Resolution Imaging

Catherine H. Kaschula[1,2], Dirk Lang[3] and M. Iqbal Parker[1,2]
[1]Department of Medical Biochemistry, University of Cape Town,
Anzio Road, Observatory, Cape Town,
[2]International Centre for Genetic Engineering and Biotechnology,
Wernher and Beit South, Anzio Rd, Observatory, Cape Town,
[3]Department of Human Biology, University of Cape Town,
Anzio Road, Observatory, Cape Town,
South Africa

1. Introduction

Fluorescence microscopy is a non-invasive technique that allows for the dynamic recording of molecular events in live cells, tissues and animals and is based on the principle that fluorescently-labelled material can be illuminated at one wavelength and emit light or fluoresce at another wavelength. The live material is selectively labelled with a fluorescent probe (the fluorophore) to generate a fluorescent image which is detected and recorded through the objective of a microscope. At the one end of the scale: positron emission tomography (PET), magnetic resonance spectroscopy (MRI) and optical coherence spectroscopy (OCT) provide real-time images from live animal or human subjects with resolutions up to about 1 mm, 100 μm and 10 μm respectively (Fernandez-Suarez and Ting 2008) (see Figure 1). At the other end of the scale: electron microscopy provides near molecular-level spatial resolution down to a few nanometers, but here cells must be fixed, which is invasive and prevents dynamic imaging. The most widely used fluorescent imaging methods in research are confocal and wide field microscopy which can provide resolutions down to a few hundred nanometers (ie can resolve intracellular organelles and track proteins in live cells). With the recent emergence of new far field fluorescence imaging techniques, it is now possible to achieve a higher level of resolution down to 10 nm to resolve single synaptic vesicles or pairs of interacting proteins (Fernandez-Suarez and Ting 2008; Huang, Babcock, and Zhuang 2010). In this chapter, we will focus on the relatively new field of far field or super resolution fluorescence imaging. The current limitations in terms of spatial and temporal resolution will be discussed together with recent fluorescent probe technology. A few applications of these techniques which have led to new discoveries will be presented.

2. The spatial resolution limit

In light microscopy, resolution is fundamentally limited by the properties of diffraction (Abbe 1873) or the "spreading out" of a light wave when it passes through a small aperture or is focused to a focal point. The diffraction barrier, which was first described by Ernst

Abbe in 1873, describes the inability of a lens-based optical microscope to discern details that are closer together than half the wavelength of light (Toomre and Bewersdorf 2010). As a result, a minutely small object that emits light will be detected as a finite-sized spot, the size of which is referred to as the point-spread function (PSF) (Figure 2). The PSF is

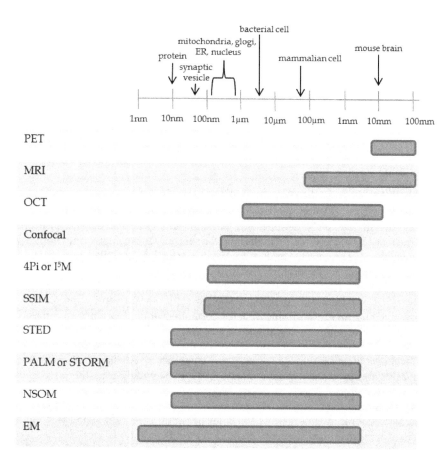

Fig. 1. **Comparison of the spatial resolutions of biological imaging techniques.** The size scale is logarithmic and approximate sizes of biological features are displayed. In addition, the spatial resolutions are estimates and are given for the focal plane. ER, endoplasmic reticulum; PET, positron-emission microscopy; MRI, magnetic resonance imaging; OCT, optical coherence tomography; SSIM, saturated structured-illumination microscopy; STED, stimulated emission depletion; PALM, photoactivated localization microscopy; STORM, stochastic optical reconstruction microscopy; NSOM, near-filed scanning optical microscopy; EM, electron microscopy. Adapted from (Fernandez-Suarez and Ting 2008).

elongated in shape along the optical axis due to the nature of the non-symmetrical wavefront that emerges from a conventional objective lens (Heilemann 2010). According to the diffraction barrier: the resolution obtainable with a wide field microscope is 200 – 250 nm in the x and y directions and 500 – 700 nm in the z-direction (Toomre and Bewersdorf 2010) which is suitable resolution to view organelles and proteins.

Imaging techniques such as multiphoton fluorescence microscopy and confocal laser microscopy have gently pushed the diffraction limit by using a focused laser beam to reduce the focal point size. In addition the confocal microscope uses a spatial pinhole to eliminate out-of-focus light thicker than the focal plane. 4Pi microscopy and I⁵M is another branch of microscopy that makes use of two opposing objective lenses to sharpen the PSF along the optical axis through interference of the counter-propagating wavefronts (Heilemann 2010). Although all of the above mentioned methods improve resolution (down to about 100 nm), they are still fundamentally limited by diffraction (Huang, Babcock, and Zhuang 2010).

Diffraction-limited resolution applies only to light that has propagated for a distance substantially larger than its wavelength (i.e. in the far field). In 1992, the first super-resolution image of a biological sample was obtained using near-field scanning optical microscopy (NSOM) (Betzig and Trautman 1992). Here the excitation source or detection probe is placed near the sample to obtain resolutions in the 20 - 120 nm range. Although NSOM has been used to study the nanoscale organisation of several membrane proteins, it cannot be used for intracellular imaging as the probe or excitation source needs to be within tens of nanometers of the target object.

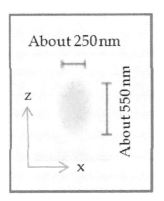
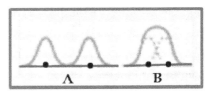

Fig. 2. **Diffraction-limited resolution of conventional light microscopy.** The focal spot of a typical objective with a high aperture is depicted by the ellispse with a width of about 250 nm in the x-direction and about 550 nm in the z-direction. The image of a point emitter imaged through the objective (the point spread function), has similar widths which define the diffraction-limited resolution. Two objects separated by a distance > the resolution limit are resolvable and appear as two separate entities in the image (i.e. in **A**) whereas images closer together than the resolution limit appear unresolvable (i.e. in **B**). Adapted from (Huang, Babcock, and Zhuang 2010).

3. Far field super resolution imaging

Features that are spectrally different are not challenged by diffraction. Likewise, Abbe's barrier does not prevent determining the coordinates of a molecule down to 1 nm with great precision (Kural et al. 2005) if there is no similar marker molecule within $\lambda/2n$ (the diffraction limited region). Overcoming the diffraction limit has been achieved by discerning groups of labelled features within a distance $<\lambda/2n$. This has been realised by modulating the emissions of fluorescent probes (i.e. transitions between bright and dark states) within a diffraction-limited region (Hell 2007).

One class of super-resolution techniques use patterned illumination to spatially modulate the fluorescence behaviour of a *population of molecules* within a diffraction-limited region, such that not all of them emit simultaneously. Microscopies utilizing this technique include stimulated emission depletion (STED), RESOLFT technology and saturated structured illumination microscopy (SSIM).

Another other class of super-resolution techniques uses photoswitching or other mechanisms to activate *individual molecules* within a diffraction-limited region. Images are then reconstructed with subdiffraction limit resolution from the measured positions of individual fluorophores. Microscopies utilizing this technique include stochastic optical reconstruction microscopy (STORM), photoactivated microscopy (PALM), and fluorescence photoactivation localization microscopy (FPALM).

4. Super-resolution imaging of a *population of molecules*

These techniques apply patterned light to a sample to manipulate its fluorescence emission. This spatial modulation can be applied in either a positive or negative manner. In the negative case, patterned light is applied to supress the population of molecules that can fluoresce. In the positive case, the light field used to excite the sample is patterned. In both of these techniques, the spatial information encoded into the illumination pattern allows neighbouring fluorophores to be distinguished from each other, leading to enhanced spatial resolution.

4.1 Principles of STED microscopy

In STED microscopy, fluorescence emission of a cluster of fluorophores is selectively "turned off" or quenched. The sample is illuminated with an excitation laser pulse which is immediately chased by a red-shifted pulse or STED beam (see Figure 3). The STED pulse quenches the fluorophores that reside within the excited state everywhere except those close to the zero intensity position to give a doughnut emission profile. When the two pulses are superimposed, only molecules close to the zero of the STED beam fluoresce, thereby lowering the PSF and increasing resolution. This approach offers improved resolution given a strong depletion light source, low scattering from the sample and good photostability of the fluorophores. In biological samples, STED images have achieved a resolution down to 20 nm in the case of organic dyes and 50 - 70 nm in the case of fluorescent proteins (Fernendez-Suarez and Ting 2008; Huang, Babcock, and Zhuang 2010).

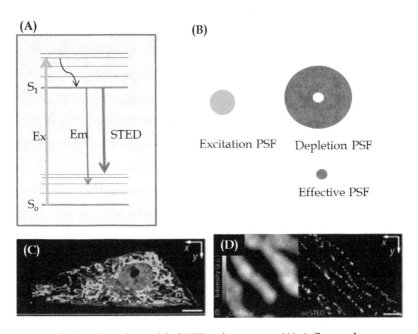

Fig. 3. **Super-resolution imaging with STED microscopy.** (A) A fluorophore can enter the first excited state S_1, following the absorption of a photon of appropriate energy. Fast relaxation to the vibrational ground state of S_1 can cause the emission of fluorescence to occur. The key principle of STED microscopy is that this excited state is locally depopulated by inducing stimulated emission. (B) A first laser that excites fluorophores into the S_1 state is overlaid with a depletion laser which has a doughnut-shaped intensity profile, where the area of zero-intensity scales with the irradiation intensity of the depletion beam. The resulting "effective" PSF represents the remaining area where fluorescence emission is still observed, and which is well below the diffraction limit. (C) Overview of the mitochondrial network of a PtK2 (kangaroo rat) cell. The mitochondria was labelled with antibodies against the translocase of the outer membrane of mitochondria (TOM) complex (green) and the microtubule cytoskeleton was labelled with antibodies against β-tubulin (red). The nucleus was stained with DAPI (blue). Scale bar 10 μm (D) Mitochondria labelled for the outer membrane with antibodies specific for the TOM complex imaged with a confocal microscope (left) and isoSTED nanoscope (right). Scale bar 500 nm. Both (C) and (D) are reprinted with permission (Schmidt et al. 2009). Copyright 2009 American Chemical Society.

4.2 Principles of SIM microscopy

SIM microscopy utilizes a positive sinusoidal pattern of excitation light by combining two light beams. A final image is computationally reconstructed from multiple snapshots collected by scanning and rotating the pattern. Spatial modulation from the excitation pattern brings about enhanced spatial resolution (see Figure 4).

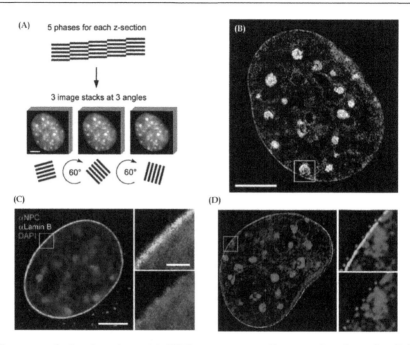

Fig. 4. **Super-resolution imaging with SIM spectroscopy.** Cross section through a DAPI-stained C2C12 cell nucleus aacquired with structured illumination. Five phases in the sine wave pattern were recorded at each z-position (in **A**), allowing the shifted components to be separated and returned to their proper location in space. Three image stacks are recorded with the diffraction grating rotated to three positions 60° apart. The cross section is reconstructed to give the 3D SIM image (in **B**). Scale bar 5 μm. (**C**) Cross section of a confocal image of the nucleus stained for DNA (blue), lamin B (green), and the nuclear pore complex (red), The right panes show the magnified images of the boxed region. (**D**) 3D SIM image of a similarly stained nucleus. Reprinted with permission (Schermelleh et al. 2008).

4.3 Video rate super-resolution images of live cells using STED and SIM

Temporal resolution refers to the precision of a measurement with respect to time, which is critical for the dynamic imaging in living cells. There is an interplay between temporal and spatial resolution due to the finite speed of light and the time taken for the photons to reach the detector. During this timeframe, the system may have undergone a change, thus the longer the light has to travel, the lower the temporal resolution. Video rate STED imaging (28 frames per second) with 62 nm spatial resolution has been demonstrated in a field of view of about 5 μM^2, allowing the motion of individual synaptic vesicles in a dendritic spine to be followed (Westphal et al. 2008). This was achieved by increasing laser intensity (to 400 mW per cm^2) and reducing the number of photons collected per imaging cycle (resulting in increased spatial resolution). This situation is not ideal as the high laser intensities not damaging to living cells. Replacing the pulsed STED lasers with continuous wave lasers permits faster scanning and higher time resolution (Willig et al. 2007). In this regard, a 70 μM^2 image of an endoplasmic reticulum took only 0.19s to acquire (Moneron et al. 2010).

SIM is good for live-cell applications that require a large field of view but not very high spatial resolution as this technique is limited by how fast the illumination pattern can be modulated and the rate of camera speed (Huang, Babcock, and Zhuang 2010).

5. Super-resolution fluorescence microscopy by *single-molecule switching*

Fluorescent probes with photoswitchable properties have been developed to modulate the fluorescence emission profile of individual fluorophores such that only an optically resolvable subset of fluorophores are activated at any moment, allowing their localization with high accuracy. Over the course of multiple activation cycles, the positions of numerous fluorophores are determined and used to construct a high-resolution image. This is the basis of PALM, FPALM and STORM super-resolution spectroscopy. Here spatial resolution is dependent on the precision of a molecules' position which in turn is related to the number of photons which are detected. For example, in the absence of background, if 10 000 photons are collected from a single fluorophore before it bleaches or is turned off, its position can be determined to 2 nm precision (Yildiz et al. 2003).

Being able to localize a single molecule does not directly translate to super-resolution imaging as the labelled biological sample may contain thousands of fluorophores within a diffraction limited region. The fluorescence emissions of the fluorophores will overlap such that the overall image will appear as a blur. However, if the fluorescence emission from these molecules is controlled such that only one molecule is emitting at a time, individual molecules can be imaged and localized.

6. Fluorescent probes used in super-resolution imaging

STORM and PALM microscopy are made possible only though the use of fluorescent probes. Despite the high specifications required for these probes, a large number of switchable fluorophores are available. These probes must firstly have a fluorescent state that emits light at one wavelength and a dark state that does not emit light at this wavelength. Secondly, in order to achieve high precision of localization, the probes should emit a large number of photons before entering the dark phase. Thirdly, because only one fluorophore is activated within a diffraction-limited area at any time, the fluorophores within the dark state should remain as such to ensure high precision localization of the activated fluorophore. A low spontaneous rate of activation of the fluorophores in the dark state is also desired (i.e. spontaneous activation by thermal energy) (Huang, Babcock, and Zhuang 2010). Currently available probes range from organic dyes to fluorescent proteins. Some of these will be discussed below.

6.1 Fluorescent proteins

There are two classes of fluorescent proteins used in super-resolution imaging: those that convert from a dark to a bright fluorescent state upon irradiation (called photo-activatable proteins), and those whose fluorescence wavelength shifts upon irradiation (also called photoshiftable fluorescent proteins). All known photoshiftable proteins shift their wavelength emission irreversibly, whereas other non-photoshiftable fluorescent proteins emit both reversibly and irreversibly (Fernandez-Suarez and Ting 2008; Lukyanov et al. 2005).

EosFP is the most commonly used irreversible photoswitchable fluorescent protein which exhibits both a high contrast and brightness (Wiedenmann et al. 2004). This protein emits strong green fluorescence (516 nm) that changes to red (581 nm) upon near UV irradiation because of a photo-induced modification involving a break in the peptide backbone next to the chromophore (see Figure 5). This protein was used successfully to perform single-particle tracking of membrane proteins in live COS7 cells at an imaging speed of 20 frames per second using PALM (Manley et al. 2008). The main disadvantage of monomeric EosFP however, is that the chromophore formation occurs only at temperatures below 30 °C, which limits its use in mammalian cells (Wiedenmann et al. 2004). Even the brightest photoswitchable fluorescent proteins are still much dimmer than many of the small molecule organic fluorophores. For example EosFP provides about 490 collected photons per molecule (Schroff et al. 2007) whereas the switchable fluorophore pair Cy3-Cy5 provides about 6000 collected photons per molecule per switching cycle which lasts about 200 cycles (Bates et al. 2007; Bates, Blosser, and Zhuang 2005).

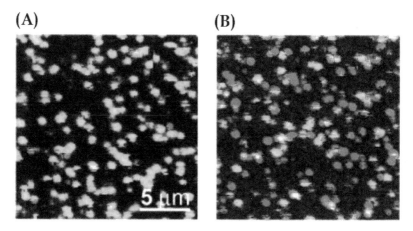

Fig. 5. **Single-molecule spectroscopy of EosFP immobilized on a BSA surface**. Confocal images were taken at 488 nm excitation (**A**) and 400 nm excitation (**B**). Reprinted with permission (Wiedenmann et al 2004)

Reversible fluorescent proteins are advantageous in super-resolution imaging as the same fluorophore can be imaged multiple times. Reversible photoswitching is a prerequisite in RESOLFT imaging, in which each molecule is switched on and off many times in order to reconstruct a subdiffraction image. The best known reversible fluorescent protein is the naturally occurring Dronpa (Ando, Mizuno, and Miyawaki 2004) and its variants of which Padron is one of them (Andresen et al. 2008).

6.2 Organic dyes

There are three main classes of non-genetically encoded probes that have been used in super-resolution imaging, namely inorganic quantum dots, reversible photoswitches and irreversible photocaged fluorophores.

Fluorescent molecules suitable for STED imaging need to have a high quantum yield and slow fluorescence decay, in which case ATTO or DY dyes are ideal. For RESOLFT imaging, the photoswitches FP595 and futyl fulgides are useful (Fernandez-Suarez and Ting 2008). The small molecule analogues to the reversible photoactivatable proteins (i.e. Dronpa) are photochromic probes which include rhodamines and diarylethenes and photoswitchable cyanines. These dyes have higher contrast ratios and higher extinction coefficients than their fluorescent protein counterparts, resulting in a larger number of photons collected per molecule. The photoswitchable cyanines have been used in both PALMIRA and STORM imaging (Bates et al. 2007; Huang et al. 2008) (see Figure 6 below). Cy5 is best used in combination with a secondary chromophore (or activator) that facilitates the switching. For example when Cy5 is paired with Cy3, the same red laser that excites Cy5 is also used to switch the dye to a stable dark state. Subsequently, exposure to green laser light converts Cy5 back to the fluorescent state, and this recovery depends on the close proximity of the secondary dye Cy3 (Bates, Blosser, and Zhuang 2005). Cy3 has also been found to facilitate switching of other cyanines which has greatly increased the amount of colours that are available for STORM imaging and has allowed for the simultaneous visualization of microtubules and clathrin-coated pits in fixed mammalian cells with 20 - 30 nm lateral resolution (see Figure 7) (Bates et al. 2007). The availability of several colours of photoswitchable cyanine dyes gives these fluorophores more diverse application than the photoswitchable fluorescent proteins of which only a few colours are available. Photoswitchable rhodamines are also an important class of photoswitches as they are membrane permeable which enables their use for live-cell imaging, compared to the cyanine dyes which are not.

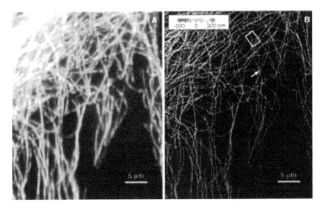

Fig. 6. **A three dimensional STORM image of microtubules in a BS-C-1 cell.**
(**A**) Conventional immunofluorescence imaging of microtubules. (**B**) The 3D STORM image of the same area using Cy3 and Alexa 647 photoswitchabe cyanine pair. A red laser (657 nm) was used to image Alexa 647 molecules and deactivate them to the dark state; a green laser (532 nm) was used to reactivate Alexa 646 in a Cy3-dependent manner. Reprinted with permission (Huang et al. 2008).

Another important class of dyes are the irreversible caged fluorophores such as the caged Q-rhodamine (Gee, Weinberg, and Kozlowski 2001) although these compounds have not been used for super-resolution imaging of biological samples.

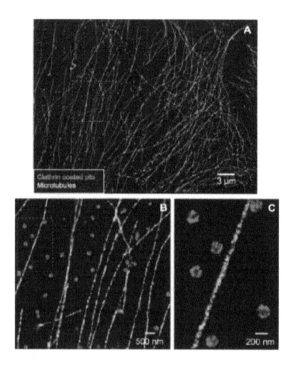

Fig. 7. **Two-colour STORM imaging of microtubules and clathrin coated pits in a mammalian cell.** (**A**) STORM image of a large area of a BS-C-1 cell. The microtubules were immunostained with Cy2 and Alexa 647, and those for clathrin with Cy3 and Alexa 647. The 457 nm and 532 nm laser pulses were used to selectively activate the two pairs of fluorophores. Each localization was false coloured according to the following code: green for 457 nm activation and red for 532 nm activation. (**B**) Enlarged STORM image of the boxed area. (**C**) Further magnification of the boxed area. Reprinted with permission (Bates et al. 2007).

7. Site-specific targeting of fluorophores to cellular proteins

Although non-genetically encoded probes generally show increased brightness and photostability compared to their fluorescent protein counterparts, they have their disadvantages. The lack of genetic encoding means that these probes require targeting to the biomolecule of interest inside the cell. These probes have been traditionally targeted using antibodies although their application is not widespread. Antibodies are not membrane permeable, and hence are not useful for labelling living cells intracellulary. Antibody staining also usually results in a low labelling efficiency and the large size of antibodies contributes to uncertainty in the spatial relationship between the label and target (Fernandez-Suarez and Ting 2008).

Some current approaches to site-specific labelling of biomolecules in living cells has been reviewed by Fernandez-Suarez and Ting (Fernandez-Suarez and Ting 2008). One method involves fusion of a peptide that recruits a small molecule to the protein of interest (Martin et al. 2005; Lata et al. 2006). Other methodologies use proteins to recruit the small molecule tag (Marks, Braun, and Nolan 2004; Bonasio et al. 2007) which can improve the specificity of binding due to the larger interaction surface although the increased size of this protein can perturb protein/enzyme function. In a combination method which seeks to achieve high labelling specificity with minimal perturbation to the protein target, a peptide recognition sequence has been used comprising an enzyme to catalyse the attachment of the probe to the sequence (Fernandez-Suarez et al. 2007).

8. Perspectives on emerging applications of super-resolution microscopy in live cells

The major technological principles of super-resolution microscopy (SIM, STED and PALM/STORM) have now matured to the extent that they have been implemented in commercially available systems that are relatively easy to use and within reach for well-established research laboratories. Thus, it is likely that we are standing at the beginning of an era of groundbreaking discoveries, fuelled by a multitude of applications of these novel imaging approaches to the challenging questions in cell biology.

Substantial potential for super-resolution imaging exists, for example, in understanding the structural basis of signal transduction within cells. Aspects of the organization and function of lipid rafts or microdomains in the cell membrane have been controversially discussed in the past, and imaging with resolution on the nanometer scale now allows addressing questions such as the molecular composition and dynamics of putative signaling complexes (Lang and Rizzoli 2010) (see Figure 8), the dynamic cytoskeletal changes underlying cell motility and migration, the way plasmamembrane structures are linked to and interact with the cytoskeleton (Ahmed 2011), or how cells interact with substrate molecules.

Our understanding of how cells communicate *in vitro* or even in the context of live tissues is set to benefit substantially from super-resolution technologies. The STED approach has been used to analyse the subcellular distribution of Na-K-ATPase in neurons (Blom et al. 2011) and to map synaptic spines in live brain tissue (Nagerl and Bonhoeffer 2010) (see Figure 9). Protein localization in chemical synapses has been investigated using STORM imaging (Dani et al. 2010). In the context of immunology, super-resolution imaging has been applied to study the composition of the immunological synapse (Dani et al. 2010) (See Figure 10) and it is now well within reach to visualize the dynamic process of how viral particles interact with immune cells, as recently shown (Felts et al. 2010). It is even possible to map GFP-tagged proteins in live multicellular organisms, as has been demonstrated in the nematode C. elegans, using STED (Rankin et al. 2011).

Super-resolution microscopy will also enhance our ability to study molecular interactions, based on signal colocalization, FRET analysis or the genetic engineering of constructs that emit fluorescence when two interaction partners are in close proximity, as has been demonstrated (Ahmed 2011).

Beyond the study of proteins, super-resolution imaging, particularly STED due to its high temporal resolution and the fact that it is based on the simultaneous imaging of a number of fluorophores in a given volume, has the potential of becoming a powerful tool to study cell physiology using diffusible fluorescent indicator dyes, e.g. for Ca^{2+} (Nagerl and Bonhoeffer 2010). Single-molecule super-resolution approaches using such dyes have been employed to visualize single ion channels (Patterson et al. 2010; Wiltgen, Smith, and Parker 2010).

In summary, each of the different approaches to super-resolution microscopy holds enormous potential in addressing key questions in current cell biology. However, they also have their characteristic advantages and drawbacks. SIM is a widefield technique easily implemented and not very demanding in terms of specimen preparation and labelling that can be used for multichannel fluorescence detection and is reasonably suitable for imaging of dynamic processes (image acquisition rates upward of 10 frames/s are possible), but it has comparatively low resolution upward of 50 nm. PALM, STORM and their derivatives are widefield fluorescence microscopy-based techniques that are currently achieving the highest resolution (in the range of 20 nm) and allow for multi-channel fluorescence imaging, but are largely confined to analysis of static or relatively slow processes in the order of minutes and in thin monolayers of cells or tissue sections. STED is a confocal laser scanning-based technique, allowing for imaging of fast dynamic processes in the range of milliseconds and analysis of relatively thick tissue slices with high lateral resolution. STED as well as PALM/STORM techniques have very specific requirements with regard to specimen preparation and labelling and the potential of these techniques is still limited to some extent by the availability of suitable fluorophores.

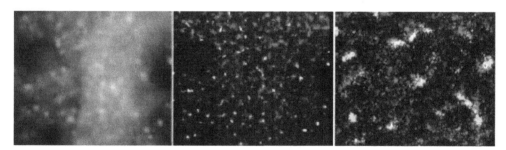

Fig. 8. **Application of TIRF and PALM imaging**. Demonstration of clusters of transferrin receptor (labeled with PalmCherry, red) and clathrin light chain (labeled with PAGFP, green) in the cell membrane by TIRF-microscopy (left) and PALM (middle; right, magnified view) (Lang and Rizzoli 2010).

As super-resolution microscopy techniques become established tools in cell biology research, a future challenge will be to design multimodal imaging approaches that combine the strengths of the different techniques. There is also a need to develop more fluorophores that are suitable for live-cell labelling, have sufficient quantum yield and provide a palette of spectral ranges suitable for the sensitive and simultaneous labelling of multiple cellular components. The drive towards a more sophisticated microscope, light source and computing hardware is still likely to lead to substantial improvements in the theoretically unlimited resolution beyond the diffraction barrier, and will enhance the capability of the systems for temporal resolution and 3-dimensional imaging.

Fig. 9. **Application of STED imaging**. STED-based 3-dimensional reconstruction of dendritic spines genetically tagged with GFP. Scale bar: 1µm (Nagerl and Bonhoeffer 2010).

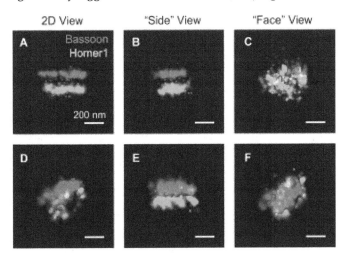

Fig. 10. **Application of STORM imaging**. STORM imaging of the pre-synaptic protein Bassoon and post-synaptic Homer1 using STORM super-resolution microscopy (Dani et al. 2010).

9. Conclusion

With the development of super-resolution imaging techniques it is now possible to image live cells down to tens of nanometers. STED imaging has allowed video rate tracking of synaptic vesicles in a dendritic spine down to 62 nm spatial resolution (Westphal et al. 2008), whereas STED, STORM and PALM have allowed cellular structures to be imaged in 3D and multiple colours. With such improved resolution, protein pairs have been visualised that contradict previous reports (Shroff et al. 2007) demonstrating the power of visualizing biomolecules of high resolution. For further improvements in spatial and temporal resolution, increased computational methods as well as fluorophores and site-specific live cell labelling are required.

10. Acknowledgments

This work was supported by the South African Research Chairs Initiative (SARCHI) of the Department of Science and Technology, the National Research Foundation (NRF), and research grants from the Medical Research Council (MRC) of South Africa and the University of Cape Town (UCT).

11. References

Abbe, E. 1873. "Beitrage zur Theorie des Mikroscops und der mikroskopischen Wahrnehmung." *Arch. Mikr. Anat.* no. 9:413-468.

Ahmed, S. 2011. "Nanoscopy of cell architecture: the actin-membrane interface." *Bioarchitecture* no. 1:32-38.

Ando, R., H. Mizuno, and A. Miyawaki. 2004. "Regulated fast nucleocytoplasmic shuttling observed by reversible protein highlighting" *Nature Biotechnology* no. 306 (5700):1370-1373.

Andresen, M., A.C. Stiel, J. Folling, D. Wenzel, A. Schonle, A. Egner, C. Eggeling, S.W. Hell, and S. Jakobs. 2008. "Photoswitchable fluorescent proteins enable monochromatic mulilabel imaging and duel color fluorescence nanoscopy." *Nature Biotechnology* no. 26 (9):1035-1040.

Bates, M., T.R. Blosser, and X. Zhuang. 2005. "Short-Range Spectroscopic Ruler Based on a Single-Molecuke Opical Switch." *Physical Review Letters* no. 94:108101.

Bates, M., B. Huang, G.T. Dempsey, and X. Zhuang. 2007. "Multicolor Super-Resolution Imaging with Photo-Switchable Fluorescent Probes." *Science* no. 317:1749-1753.

Betzig, E., and J.K. Trautman. 1992. "Near-field optics: microscopy, spectroscopy, and surface modification beyond the diffraction limit." *Science* no. 257 (5067):189-195.

Blom, H., D. Ronnlund, L. Scott, Z. Spicarova, J. Widengren, A. Bondar, A. Aperia, and H. Brismar. 2011. "Spatial distribution of Na+ K+ ATPase in dendritic spines dissected by nanoscale superresolution STED microscopy." *BMC Neuroscience* no. 12:16.

Bonasio, R., C.V. Carman, E. Kim, P.T. Sage, K.R. Love, T.R. Mempel, T.A. Springer, and U.H. von Andrian. 2007. "Specific and covalent labeling of a membrane protein with organic fluorochromes and quantum dots." *Proceedings of the National Academy of Science USA* no. 104 (37):14753-14758.

Dani, A., B. Huang, J. Bergan, C. Dulac, and X. Zhuang. 2010. "Superresolution imaging of chemical synapses in the brain." *Neuron* no. 68 (5):843-856.

Felts, R.L., K. Narayan, J.D. Estes, D. Shi, C.M. Trubey, J. Fu, L.M. Hartnell, G.T. Ruthel, D.K. Schneider, K. Nagashima, J.W. Jr. Bess, S. Bavari, B.C. Lowekamp, D. Bliss, J.D. Lifson, and S. Subramaniam. 2010. "3D visualization of HIV transfer at the virological synapse between dendritic cells and T cells." *Proceedings of the National Academy of Science USA* no. 107 (30):13336-13341.

Fernandez-Suarez, M., H. Baruah, L. Martinez-Hernandez, K.T. Xie, J.M. Baskin, C.R. Bertozzi, and A.Y. Ting. 2007. "Redirecting lipoic acid ligase for cell surface protein labeling wioth small molecule probes." *Nature Biotechnology* no. 25 (12):1483-1487.

Fernandez-Suarez, M., and A.Y. Ting. 2008. "Fluorescent probes for super-resolution imaging in living cells." *Nature Reviews Molecular Cell Biology* no. 9:292-944.

Gee, K.R., E.S. Weinberg, and D.J. Kozlowski. 2001. "Caged Q-rhodamine dextran: a new photoactivated fluorescent tracer." *Bioorganic and Medicinal Chemistry Letters* no. 11:2181-2183.

Heilemann, M. 2010. "Fluorescence Microscopy Beyond the Diffraction Limit." *Journal of Biotechnology* no. 149:243-251.

Hell, S.W. 2007. "Far-Field Optical Nanoscopy." *Science* no. 316:1153-1158.

Huang, B., H. Babcock, and X. Zhuang. 2010. "Breaking the Diffraction Barrier: Super-Resolution Imaging of Cells." *Cell* no. 143 (7):1047-1058.

Huang, B., W. Wang, M. Bates, and X. Zhuang. 2008. "Three-dimensional Super-Resolution Imnaging by Stochastic Optical Reconstruction Microscopy." *Science* no. 319:810-813.

Kural, C., H. Kim, S. Syed, G. Goshima, V.I. Gelfand, and P.R. Selvin. 2005. "Keneisin and SDynein Move a Peroxisome in Vivo: A Tug-of-War or Coordinated Movement." *Science* no. 308:1469-1472.

Lang, T., and S.O. Rizzoli. 2010. "Membrane protein clusters at nanoscale resolution: more than pretty pictures." *Physiology* no. 25 (2):116-124.

Lata, S., M. Gavutis, R. Tampe, and J. Piehler. 2006. "Specific and stable fluorescence labeling of histidibne-tagged proteins for dissecting mulit-protein complex formation " *Journal of the American Chemical Society* no. 128:2365-2372.

Lukyanov, K.A., D.M. Chudakov, S. Lukyanov, and V.V. Verkusha. 2005. "Photoactivatable Fluorescent Proteins." *Nature Reviews: Moleculaer Cell Biology* no. 6:885-891.

Manley, S., J.M. Gillette, G.H. Patterson, H. Schroff, H.F. Hess, E. Betzig, and J. Lippincott-Schwartz. 2008. "High-density mapping of Single Molecule Trajectories with Photoactivated Localization Microscopy." *Nature Methods* no. 5 (2):155-157.

Marks, K.M., P.D. Braun, and G.P. Nolan. 2004. "A general approach for chemical labelling and rapid spatially controlled protein inactivation." *Proceedings of the National Academy of Science USA* no. 101:9982-9987.

Martin, B.R., B.N. Giepmans, S.R. Adams, and R.Y. Tsien. 2005. "Mammalian cell-based optimization of the biarsenical-binding tetracysteine motif for improved fluorescence and affinity." *Nature Biotechnology* no. 23 (10):1308-1314.

Moneron, G., R. Medda, B. Hein, A. Giske, V. Westphal, and S.W. Hell. 2010. "Fast STED Microscopy with Continuous Wave Fiber Lasers." *Optics Express* no. 18 (2):1302-1308.

Nagerl, U.V., and T. Bonhoeffer. 2010. "Imaging Living Synapses at the Nanoscale by STED Microscopy." *The Journal of Neuroscience* no. 30 (28):9341-9346.

Patterson, G., M. Davidson, S. Manley, and J. Lippincott-Schwartz. 2010. "Superresolution imaging using single-molecule localization." *Annual Reviews in Physical Chemistry* no. 61:345-367.

Rankin, B.R., G. Moneron, C.A. Wurm, J.C. Nelson, A. Walter, D. Schwarzer, J. Schroeder, D.A. Colon-Ramos, and S.W. Hell. 2011. "Nanoscopy in a living multicellular organism expressing GFP." *Biophysical Journal* no. 100 (12):L63-65.

Schermelleh, L., P.M. Carlton, S. Haase, L. Shoa, L. Winito, P. Kner, B. Burke, M.C. Cardoso, D.A. Agard, M.G.L. Gustafsson, H. Leonhardt, and J.W. Sedat. 2008.

"Subdiffraction multicolor imaging of the nuclear periphery with 3D structured illumination microscopy." *Science* no. 320:1332-1336.

Schmidt, R., C.A. Wurm, C.A. Punge, A. Egner, S. Jakobs, and S.W. Hell. 2009. "Mitochondrial Cristae Revealed with Focused Light." *Nano Letters* no. 9 (6):2508-2510.

Shroff, H., C.G. Galbraith, J.A. Galbraith, H. White, J.M. Gillette, S. Olenych, M.W. Davidson, and E. Betzig. 2007. "Dual-color superresolution imaging of genetically expressed probes within individual adhesion complexes." *PNAS* no. 104 (51):20308-20313.

Toomre, D., and J. Bewersdorf. 2010. "A New Wave of Cellular Imaging." *The Annual Review of Cell and Developmental Biology* no. 26:285-314.

Westphal, V., S.O. Rizzoli, M.A. Lauterbach, D. Kamin, R. Jahn, and S.W. Hell. 2008. "Video-Rate Far-Field Optical Nanoscopy Dissects Synaptic Vesicle Movement." *Science* no. 320:246-249.

Wiedenmann, J., S. Ivanchenko, F. Oswald, F. Schmidtt, C. Rocker, A. Salih, K-D. Spindler, and G.U. Nienhaus. 2004. "EosFP, a fluorescent marker protein with UV-inducible green-to-red fluorescence conversion." *PNAS* no. 101 (45):15905-15910.

Willig, K.I., B. Harke, R. Medda, and S.W. Hell. 2007. "STED Microscopy with Continuous Wave Beams." *Nature Methods* no. 4 (11):915-918.

Wiltgen, S.M., I.F. Smith, and I. Parker. 2010. "Superresolution ,localization of single functional IP3R channels utilizing Ca2+ flux as a readout." *Biophysical Journal* no. 99 (2):437-446.

Yildiz, A., J.N. Forkey, S.A. McKinney, H. Taekjip, Y.E. Goldman, and P.R. Selvin. 2003. "Myosin V Walks Hand-Over-Hand: Singled Flurophore Imaging with 1.5nm Localization." *Science* no. 300 (5628):2061-2065.

2

One-by-One Sample Preparation Method for Protein Network Analysis

Shun-Ichiro Iemura and Tohru Natsume
Biomedicinal Information Research Center (BIRC),
National Institute of Advanced Industrial Science and Technology (AIST)
Japan

1. Introduction

Proteomics is the large-scale study of an organism's complete complement of proteins, and its relevant technologies have matured over recent years. Along with the development of mass spectrometry (MS), MS-based proteomics has emerged as an invaluable tool for large-scale identification and quantification of protein networks (Aebersold & Mann, 2003; Domon & Aebersold, 2006). Proteomic data is important for a wide range of research in basic and medical biology. In recent years, many large-scale projects have been performed and a huge amount of data has accumulated. However, because the data sets from individual projects often vary in quality, the value of proteomics for the wider scientific community is limited (Olsen & Mann, 2011).

One of the causes of this variation in proteomic data quality is thought to be the manual process of large-scale sample preparation. The sample preparation process for proteomic analysis consists of the several complicated steps. For example, sample preparation for protein interaction analysis using mammalian cells expressing a target protein typically requires 1×10^7-10^8 cells (one 10-cm or 15-cm tissue culture dish) (Blagoev et al., 2003; Burckstummer et al., 2006; Ewing et al., 2007). After cell recovery, steps such as cell lysis, purification of protein complexes, denaturation and modification of proteins, separation by gel electrophoresis, and enzymatic digestion are performed sequentially. In fact, many researchers and technicians are involved in laborious, repetitive work of large-scale sample preparation, in which they must handle tens of culture dishes at a time. In such a 'parallel sample preparation' process, during the preparation of a number of samples, the conditions undoubtedly differ between the first and last treated samples. Denaturation of the component proteins of complexes and proteolysis progress over time, and the denatured proteins are thought to be the cause of nonspecific binding. We came to realize that highly sensitive analysis could not be performed using the prevailing parallel sample preparation methods.

To optimize sample preparation conditions and improve sample quality, we considered that a 'one-by-one sample preparation' method would be useful. One-by-one sample preparation is the concept that one sample is finished at a time, followed by preparation of the next sample (Fig. 1). In this way, each sample can be prepared carefully under almost equal conditions; however, this method is not realistic for large-scale analysis, because of the large amount of human time and work involved.

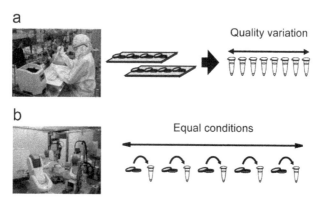

Fig. 1. Comparison of sample preparation processes. (a) Parallel preparation by the manual method. The quality of the samples was uneven. (b) One-by-one preparation. This method enables the preparation of samples under the same conditions.

To realize the one-by-one concept and perform a pilot feasibility study, a fully automated sample preparation system is required. However, in the proteomics field, partial automation for parallel preparation is usually only applied to save analysis time, to eliminate sample contamination, and to reduce human error (Alterovitz et al., 2006). Several semi-automated robots that are specialized in certain processes are commercially available, such as liquid dispenser robots, cell culture robots, and electrophoresis gel cutting robots. However, to develop a fully automated and highly precise system for sample preparation using commercial robots would be difficult, because these robots do not meet our specifications, or if they do, the integration of the robots from different vendors may prove difficult. Furthermore, robots for other multiple sample preparation processes have not yet been developed. To achieve a significant breakthrough, we need a versatile robotic system. Recently, high-performance and reliable multi-axis articulated vertical robots have been developed, and are used in various fields, such as the motor industry. The motion of these industrial robots is fast, precise, and flexible. Moreover, these robots are relatively easy to integrate with other robots and equipment. Although the robotic system requires considerable effort and patience to set up (Blow, 2008), once one of the designated conditions is determined, it becomes applicable in many other situations.

In this chapter, we assess the one-by-one sample preparation method compared with parallel preparation in protein network analysis, using an automated sample preparation system for liquid chromatography-tandem mass spectrometry (LC-MS/MS). This automated system is compatible with the single-step affinity purification technique using the Flag-tag system (Einhauer & Jungbauer, 2001), without sodium dodecylsulfate polyacrylamide gel electrophoresis (SDS-PAGE) separation. Affinity-purification is a technique for purification of physiological protein complexes using target proteins (bait proteins) fused with affinity tags, such as short epitope peptides (e.g., Flag and Myc) or tandem-affinity purification (TAP) tags (Kocher & Superti-Furga, 2007). The bait proteins are overexpressed in cells and are separated, together with the protein complexes, using affinity beads that bind to the tags. Finally, all component proteins are identified by LC-MS/MS. Using this system, we tested two Wnt signaling pathway (Rao & Kuhl, 2010) proteins, β-catenin and Axin1, as baits, and demonstrated that the one-by-one purification method using this system is highly sensitive and reproducible compared with the manual parallel purification method. The results indicate that gentle and equal preparation conditions are important for generating reliable data for large-scale protein-protein interaction network and for quantitative analysis.

2. Experimental procedures

2.1 Design and development of a robotic system for one-by-one sample preparation

The robotic system was manufactured using four 6-axis robots, FC03N (Kawasaki Heavy Industries, Ltd., Hyogo, Japan) and a 3-axis robot comprising three single-axis robots (IAI corporation, Shizuoka, Japan), with help from the Japan Support System, Co., Ltd. (Ibaraki, Japan) and Nikkyo Technos, Co., Ltd. (Tokyo, Japan). In low femtomole level analysis, the key to obtaining reliable data quickly is to minimize contaminants, such as chemicals, airborne particles, and keratin proteins. Chemicals cause background noise, which limit the sensitivity of MS by decreasing the signal to noise ratio (S/N). Airborne particles, including dust, cause the blockage of the flow path and the nano LC column. Keratin proteins also cause background noise, which disturbs the detection of low abundance of proteins. Therefore, because we needed to perform sample preparation in a super clean room, our automated robotic system was designed for clean room specification (ISO class 4).

2.2 Immobilization of Anti-Flag antibodies to magnetic beads

Anti-Flag M2 antibodies (Sigma-Aldrich, St. Louis, MO) were immobilized via covalent binding of the primary amine group with 1-Ethyl-3-[3-dimethylaminopropyl] carbodiimide hydrochloride (EDC; Thermo Fisher Scientific, Waltham, MA)–modified Magnosphere MS300 magnetic beads (JSR, Tokyo, Japan). The beads (10 mg) suspension was transferred into a 1.5 ml-microtube. The beads were washed twice with 1 ml of activation buffer (0.1 M 2-[N-morpholino]ethane sulfonic acid (MES), pH 6.0, 0.5 M NaCl) and were resuspended in 1 ml of activation buffer. EDC and N-hydroxysulfosuccinimide (Sulfo-NHS; Thermo Fisher Scientific, Waltham, MA) were then added to the beads suspension. The final concentrations of EDC and sulfo-NHS were 2 and 5 mM, respectively. The mixture was incubated for 15 min at room temperature (RT), placed on the magnet, and the supernatant was discarded. The antibody (100 μg/ml) in conjugation buffer (50 mM sodium phosphate, pH 7.4, 0.15 M NaCl) was added to the beads and the mixture was incubated for 3 hr at 4 °C. After incubation, the supernatant was discarded and quenching buffer (20 mM HEPES-NaOH, pH 7.5, 0.15 M NaCl, 50 mM ethanolamine) was added. After quenching for 2 hr at 4 °C, the beads were washed three times with 1 ml of washing buffer (50 mM Tris-HCl, pH 8.0, 0.5 M NaCl, 0.1% Triton X-100) and twice with storage buffer (20 mM HEPES-NaOH, pH 7.5, 0.15 M NaCl, 0.5% digitonin). The antibody-immobilized beads were stored in 1 ml of storage buffer at 4 °C.

2.3 Cell culture and transfection

HEK293T cells (approximately 5.0×10^6 cells per 10-cm dish) were seeded in Dulbecco's modified Eagle's medium (DMEM; Invitrogen, San Diego, CA) containing 10% heat-inactivated fetal bovine serum (FBS; Invitrogen) the day before transfection. The cells were transfected with human β-catenin or human Axin1 cDNA, using Lipofectamine 2000 (Invitrogen) according to the manufacturer's protocol. The cells were collected 24 h after transfection.

2.4 Cell collection and lysis

The culture medium was discarded from the 10-cm dish, and the HEK293T cells expressing a bait protein were scraped into 1 ml of cold phosphate buffered saline (PBS) and transferred into a 1.5 ml-microtube. After centrifugation at low speed (3,000 rpm) for 1 min at 4 °C, the supernatant was discarded, and 1.0 ml of lysis buffer (20 mM HEPES, pH 7.5, 150 mM NaCl, 50 mM NaF, 1 mM Na_3VO_4, 0.5% digitonin, 1 mM $MgCl_2$, 1 mM PMSF, 5 µg/ml leupeptin, 5 µg/ml aprotinin and 3 µg/ml pepstatin A) was added. The cells were lysed by gently mixing for a short time with a vortex mixer (parallel method) or with a pipette tip (one-by-one method). In this step, we chose the vortexing in the parallel method because we thought, in reality, this way had to be adopted in large-scale sample treatment. The lysate was centrifuged at high speed (15,000 rpm) for 10 min at 4 °C, and the cleared lysate was transferred into a microtube containing the anti-Flag antibody immobilized magnetic beads.

2.5 Immunoprecipitation

The supernatant was incubated with the magnetic beads at 4 °C for 10 min with a rotator (parallel method) or the 6-axis robot (one-by-one method; 10 times mixing → interval: 4 min at 4 °C → 10 times mixing → interval: 4 min at 4 °C). After incubation, the beads were washed twice with 1 ml of wash buffer (10 mM HEPES, pH 7.5, 150 mM NaCl, 0.1% Triton X-100). The protein complexes containing the bait protein were then mixed with 100 µl of Flag peptide (0.5 mg/ml, SIGMA) in wash buffer for 5 min at 4 °C using a mixer (parallel method) or a 'protein complexes elution device' (Fig. 2a) (one-by-one method). The eluted fraction was transferred to a new microtube.

2.6 Limited proteolysis with lysyl endopeptidase C (Lys-C)

To concentrate the purified proteins and to exchange the buffer, trichloroacetic acid (TCA) precipitation was performed. Sodium deoxycholate (DOC) was added to a final concentration of 0.1%. After mixing, TCA was added to a final 10% concentration and the solution was precipitated at 0 °C for 30 min. The protein precipitate was collected by centrifugation (15,000 rpm for 10 min at 4 °C). The supernatant was carefully removed, 1 ml of acetone (precooled at -30 °C) was added to the pellet, and vortexing was carried out until the pellet became unstuck from the bottom of the tube. The proteins were collected by centrifugation (15,000 rpm for 5 min at 4 °C) and the supernatant was removed. The pellet was redissolved in 10 µl extraction buffer (0.1 M Tris-HCl, pH 8.8, 0.05% n-octyl glucopyranoside, 7M guanidine hydrochloride) using the microtube mixer. After the proteins were dissolved almost completely, 40 µl of digestion buffer (0.1 M Tris-HCl, pH 8.8,

0.05% n-octyl glucopyranoside) was added and mixed. Finally, 0.1 µg of lysyl endopeptidase (Lys-C; Wako, Osaka, Japan) was added and the mixture was incubated over night at 37 °C.

2.7 Western blotting

HEK293T cells were transfected with human β-catenin or human Axin1 cDNA, or pcDNA3 vector (as a negative control) as described in section 2.3. The purified proteins (from the immunoprecipitation step, section 2.5) were separated by electrophoresis on 10% SDS-PAGE and transferred onto Polyvinylidene difluoride (PVDF) membranes. The membranes were blocked with 2% BSA in TBS-T for 1 h at RT, followed by incubation with each primary antibody for 1 h at RT. After incubation with the secondary antibody for 1 h at RT, protein bands were detected with an ECL detection kit.

2.8 Direct nanoflow liquid chromatography tandem mass spectrometry system (DNLC-MS/MS)

All samples were diluted 10-fold with 0.1% formic acid and analyzed (2 µl) by DNLC system (Natsume et al., 2002) coupled to a QSTAR XL (AB Sciex, Foster City, CA). Peptides were separated on a C18 reversed-phase column packed with Mightysil C18 (particle size 3 µm; Kanto Chemical, Tokyo, Japan) at a flow rate of 100 nl/min by a 40-min linear gradient from 5% to 40% acetonitrile in 0.1% formic acid, and were sprayed on-line to the mass spectrometer. MS and MS/MS spectra were obtained in an Information Dependent Acquisition (IDA) mode. Up to two precursor ions above the intensity threshold of 50 counts with a charge state from 2 to 3 were selected for MS/MS analyses (1.0 sec) from each survey scan (0.5 sec). The MS and MS/MS scan ranges were m/z 400-1500 and 100-1500, respectively.

2.9 Data analysis

Peak lists were created by scripts of Analyst QS 1.1 Software (AB Sciex) using the following parameters: 0.1 amu Mass tolerance for combining MS/MS spectra, 2 cps MS/MS export threshold, 5 Minimum number of MS/MS ions for export, 50% Centroid height percentage, and 0.05 amu Centroid merge distance. All MS/MS spectra were queried against the National Center for Biotechnology Information (NCBI) non-redundant database (human; January 25, 2011; 137,349 sequences) using an in-house Mascot server (version 2.2.1; Matrix Science, London, UK). Search parameters were as follows: MS and MS/MS tolerance of 250 ppm and 0.5 Da, respectively; enzymatic specificity allowing for 1 missed cleavage site and K cleavage (enzyme: Lys-C/P); no fixed modification; and variable modification of N-acetyl (protein N terminus) and phosphorylations (Ser, Thr, and Tyr). Proteins that were identified by two or more peptides with a peptide expectation value of $p < 0.05$ were considered as reliable identifications.

3. Results

3.1 Automated robotic system for one-by-one sample preparation

To perform precise one-by-one sample purification for protein network analysis, we designed and developed a robotic system for fully automated sample preparation from cell collection to limited proteolysis with Lys-C. This system consists of four 6-axis industrial robots, one 3-axis robot, high- and low-speed centrifuges, a CO_2 incubator, and other components, as illustrated in detail in Fig. 2.

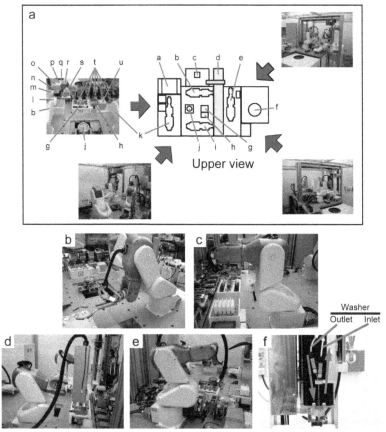

Fig. 2. Layout of the automated robotics for one-by-one sample preparation system.
(a) A schematic upper view diagram of the system and four photographs showing different views indicated by arrows. a: CO_2 incubator; b: 6-axis robot No. 2; c: low-speed centrifuge; d: 3-axis robot; e: 6-axis robot No. 4; f: high-speed centrifuge; g: microtube carriers for low-speed centrifuge; h: buffers position (lysis buffer and phosphate-buffered saline); i: 6-axis robot No. 3; j: culture dish stage; k: 6-axis robot No. 1; l: cell scrapers specialized for this system; m: pipette tips (2-200 µl); n: pipette tips (0.1-10 µl); o: protein complexes elution device; p: incubator (4 °C); q: microtube rack; r: incubator (37 °C); s: reagents rack (elution buffer, TCA, etc.); t: microtube capper/decapper (temperature-controlled); u: pipette tip (200-1,000 µl). (b) 6-axis robot No. 1: culture dish-carrying robot. (c) 6-axis robot No. 2: scraping and tube-carrying robot. (d) 6-axis robot No. 3: dispenser robot. (e) 6-axis robot No. 4: microtube-carrying robot. (f) 3-axis robot: micro-dispenser robot. A washer is attached to this robot.

The features of this system are: (i) The system is optimized for sample preparation from 10-cm culture dishes, and the process operates under gentle conditions to decrease protein denaturation and degradation compared to manual operation. The scraping robot (6-axis robot No. 2) can collect cells gently in a single scraping motion (Fig. 3a and 3b). In addition, a microtube delivery robot (6-axis robot No. 4) can mix the magnetic beads immobilized on the anti-Flag M2 antibody with cell extracts at intervals that will not over-mix or create a foam. Moreover, the elution of the protein complexes in the 'protein complexes elution device' (Fig. 2a) is performed by moving the beads backwards and forwards in the elution buffer between two magnets (Fig. 3c-e). The solution is not mixed vigorously; therefore, this procedure is expected to prevent the denaturation of the eluted protein. (ii) This system allows rapid purification of the protein complexes. One sample, from cell scraping to elution of protein complexes, can be prepared in 40 min. The manual parallel treatment of 20 samples takes more than 120 min. (iii) The one-by-one system can operate 24 hours a day, automatically, generating approximately 500 samples per month.

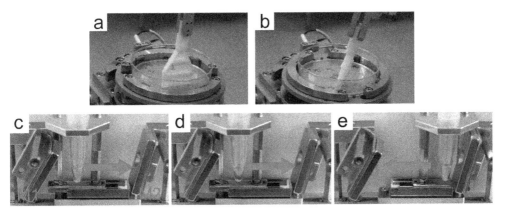

Fig. 3. Automated one-by-one sample preparation system. (a and b) Cell collection on the dish stage. (c-e) Process for elution of the protein complexes in the 'protein complexes elution device' (Fig. 1a). M1 and M2: magnets.

3.2 Comparison of parallel and one-by-one methods for the sample preparation by western blot analysis

To evaluate one-by-one sample preparation, we chose β-catenin and Axin1 as bait proteins because they are well-studied proteins that play key roles in the Wnt signaling pathway, and because, to date, many partners that interact with them have been identified (Daugherty & Gottardi, 2007; H. Huang & He, 2008; S.M. Huang et al., 2009). Furthermore, it is difficult to analyze β-catenin and Axin1-interacting proteins using affinity purification and LC-MS/MS, because these bait proteins are likely to be degraded, not only by the ubiquitin-proteasome system, but also nonspecifically by various proteases during the purification steps, even if protease inhibitors are added. Therefore, we expected that the gentle one-by-one purification method would allow these proteins to remain intact to the greatest extent possible, and would permit the identification of more interacting partner proteins.

We first compared the bait proteins (β-catenin and Axin1) from parallel preparation with those of one-by-one preparation. Flag-tagged β-catenin or Axin1 was expressed in HEK293T cells, purified by the parallel and the one-by-one method, and analyzed by western blotting (Fig. 4). In the case of parallel preparation, both β-catenin and Axin1 were found to be degraded. In particular, Axin1 degradation tended to be fast, and a protein band of approximately 120 kDa, corresponding to the intact form, was almost absent in some cases. On the other hand, in samples prepared by the one-by-one method, degradation of the bait proteins was significantly reduced. Interestingly, for Axin1, only one prominent band of the size of the intact protein was detected in most cases. These data indicated that the one-by-one method minimizes protein denaturation and degradation during sample preparation compared to the parallel method.

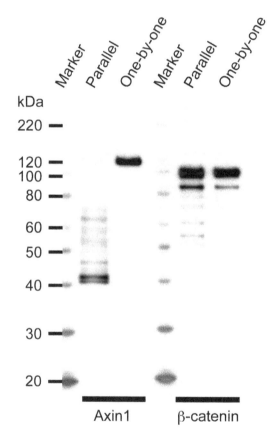

Fig. 4. Comparison of bait protein (β-catenin and Axin1) purification quality. Flag-tagged β-catenin or Axin1 proteins were expressed in HEK293T cells, purified by the parallel or one-by-one methods until the elution steps, and analyzed by western blot analysis. One-by-one: automated one-by-one method; Parallel: manual parallel method.

3.3 Comparison of parallel and one-by-one methods for the sample preparation by protein network analysis

Next, we compared the component proteins interacting with the bait proteins (β-catenin and Axin1) prepared by the parallel and one-by-one methods. Each bait protein was expressed in HEK293T cells and purified with its binding partner proteins. These proteins were then digested with Lys-C and analyzed by a DNLC-MS/MS system (Natsume et al., 2002). The identified proteins that interact with β-catenin and Axin1, excluding nonspecific binding, are listed in Table 1. As expected, the one-by-one preparation method showed better detection sensitivity and reproducibility compared with the parallel method.

Bait: β-catenin

Name[a]	Symbol[a]	Parallel[b] reproducibility (n = 10)	One-by-one[b] reproducibility (n = 10)
Adenomatous polyposis coli	APC	10 (100%)	10 (100%)
Adenomatosis polyposis coli 2	APC2	0	8 (80%)
Axin 1	AXIN1	10 (100%)	10 (100%)
Axin 2	AXIN2	1 (10%)	9 (90%)
Beta-transducin repeat containing	BTRC | FBXW11	1 (10%)	10 (100%)
Cadherin 1, type 1	CDH1	1 (10%)	9 (90%)
Cadherin 2, type 1	CDH2	2 (20%)	10 (100%)
Casein kinase 1, alpha 1	CSNK1A1	10 (100%)	10 (100%)
Catenin, alpha 1	CTNNA1	10 (100%)	10 (100%)
Catenin, alpha; 1 or 2	CTNNA1 | CTNNA2	10 (100%)	10 (100%)
Catenin, alpha; 1 or 3	CTNNA1 | CTNNA3	10 (100%)	10 (100%)
Catenin, beta interacting protein 1	CTNNBIP1	8 (80%)	10 (100%)
Catenin, delta 1; isoform 1B	CTNND1	0	10 (100%)
Cathepsin A	CTSA	9 (90%)	10 (100%)
Cullin 1	CUL1	2 (20%)	10 (100%)
Ezrin	EZR	0	8 (80%)
Family with sequence similarity 123B	FAM123B	10 (100%)	10 (100%)
F-box and WD repeat domain containing 11	FBXW11	1 (10%)	10 (100%)
Galactosidase, beta 1	GLB1	8 (80%)	10 (100%)
Glycogen synthase kinase 3 alpha	GSK3A	2 (20%)	10 (100%)
Glycogen synthase kinase 3; alpha or beta	GSK3A | GSK3B	6 (60%)	10 (100%)
Glycogen synthase kinase 3 beta	GSK3B	4 (40%)	10 (100%)
HMG-box transcription factor TCF-3	TCF7L1	7 (70%)	10 (100%)
Lymphoid enhancer-binding factor 1	LEF1	8 (80%)	10 (100%)

Bait: β-catenin

Name[a]	Symbol[a]	Parallel[b] reproducibility (n = 10)	One-by-one[b] reproducibility (n = 10)
S-phase kinase-associated protein 1	SKP1	2 (20%)	10 (100%)
Transcription factor 7 (T-cell-specific, HMG-box); isoform 1	TCF7	3 (30%)	10 (100%)
Transcription factor 7-like 2	TCF7L2	10 (100%)	10 (100%)

Bait: Axin1

Name[a]	Symbol[a]	Parallel[b] reproducibility (n = 10)	One-by-one[b] reproducibility (n = 10)
Adenomatous polyposis coli	APC	0	10 (100%)
Beta-catenin	CTNNB1	0	10 (100%)
Casein kinase 1, alpha 1	CSNK1A1	1 (10%)	10 (100%)
Glycogen synthase kinase 3 beta	GSK3B	2 (20%)	10 (100%)
Macrophage erythroblast attacher	MAEA	0	10 (100%)
WD repeat domain 26; isoform b	WDR26	0	8 (80%)

Table 1. Comparison of identified proteins and their reproducibility from samples prepared by parallel and one-by-one methods (analyzed by MS). [a]Protein names and Symbols refer to the Entrez Gene database. The proteins identified by a common peptide sequence are indicated by 'or' in the Name column, and ' | ' in the Symbol column. The identified proteins exclude nonspecific proteins (Table 2). [b]The samples were prepared independently by the parallel or the one-by-one method and analyzed by the DNLS-MS/MS system.

In the analysis of the one-by-one preparation β-catenin, we identified membrane proteins (Cadherins 1 and 2), peripheral membrane proteins (δ-catenin and Ezrin), the Skp1- Cullin-F-box-protein (SCF) E3 ubiquitin ligase complex (BTRC/FBXW11, Skp1, and Cullin1) and other component proteins (Adenomatosis polyposis coli 2 (APC2) and Axin2) using the one-by-one method, whereas some of these proteins were not identified by the parallel method. The reproducibility increased from below 20% (parallel preparation, n = 10) to above 80% (one-by-one preparation, n = 10). In the analysis of Axin1, the one-by-one method dramatically increased the precision of the identification of well-known interaction partners, such as Adenomatous polyposis coli (APC), δ-catenin, Glycogen synthase kinase 3β (GSK3β), and Casein kinase 1, whereas no specific interactions were identified using the parallel method (Table 1). This improvement is probably the result of the minimal degradation of Axin1 (Fig. 4). Furthermore, we found two new interacting partners: MAEA and WDR26. To confirm these interactions, Flag-tagged Axin1 was expressed in HEK293T cells and the cell extracts were subjected to immunoprecipitation with anti-Flag antibody, followed by western blotting with anti-MEAE or anti-WDR26 antibody. As shown in Fig. 5, both MAEA and WDR26 were found to form a complex with Axin1. Further work is required to determine the biological relevance of these interactions.

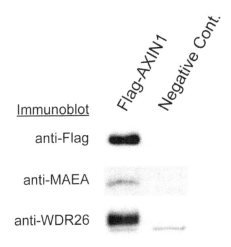

Fig. 5. Interaction of Axin1 with MAEA and WDR26. HEK293T cells were transfected with Flag-Axin1 or an empty vector (pcDNA3) as a negative control (Negative cont.). Expressed protein complexes were purified by the automated one-by-one methods until the elution step and analyzed by western blot analysis.

4. Discussion

Sample preparation is one of the most important processes for MS-based proteomics, such as large-scale protein-protein interaction networks and quantitative analyses. In affinity purification, although the single Flag-tag purification MS approach is useful and raises the possibility of identification of low abundant and transient interacting proteins, the problem is that this approach leads to a high false positive rate (Chen & Gingras, 2007). To overcome this problem, several protocols have been devised (Burckstummer et al., 2006; Selbach & Mann, 2006), and computational data processing to remove nonspecific proteins is performed during large-scale analysis (Ewing et al., 2007; Ho et al., 2002; Gavin et al., 2002). However, because it is possible to reliably identify low amounts of true interacting proteins by improving the signal-to-noise ratio in LC-MS/MS, we considered that reproducibly decreasing the level of nonspecific noise proteins in single-step purification samples would be a valid approach. Therefore, we empirically developed and optimized the conditions for sample preparation, and using this methodology, found more than fifty significant protein-protein interactions (Hirano et al., 2005; Kitajima et al., 2006; Iioka et al., 2007; Komatsu et al., 2007; Nishiyama et al., 2009; Kaneko et al., 2009; Komatsu et al., 2010). In spite of this useful methodology, we realized the limitations of the existing preparation method in large-scale analysis, because we found that the amount of true interactors, as well as nonspecific proteins, in manually parallel-prepared samples varied. The ultimate solution for this problem was to use a one-by-one purification method. In addition, because this preparation process needs to be automated to prepare samples under precisely equal conditions, we designed and developed a fully automated robotic sample preparation system for LC-MS/MS.

In a validation study using the Wnt signaling pathway proteins, β-catenin and Axin1, the rate of protein degradation was significantly higher in the parallel preparation compared with the one-by-one preparation. This higher protein degradation in parallel preparation is probably caused by the manual scraping of cells and increased preparation time. In parallel preparation, manual scraping of cells involves several rapid strokes, which may increase the cells' susceptibility to damage and increase the level of proteolytic enzymes released from subcellular compartments. The proteases, similarly to nonspecific binding proteins, are likely to attach to and degrade the purified protein complexes over time, and these degraded and denatured proteins are thought to cause nonspecific binding.

In contrast to manual parallel preparation, an important feature of the one-by-one system is the careful and brief sample preparation. The association rate of nonspecific proteins is thought to be slower than that of specific binding proteins; therefore, the careful and rapid one-by-one method reduces nonspecific protein associations. In fact, as shown in Table 2, the number of nonspecific proteins precipitated using the one-by-one method was significantly lower than that by the parallel method. Using the one-by-one method, this decrease was accompanied by a remarkable increase in known interactors, because the signal-to-noise ratio was increased in combination with the prevention of protein degradation. Although it was previously reported that single-affinity tag purifications brought an increase in nonspecific binding proteins (Chen & Gingras, 2007), we have found that the single-step one-by-one purification using anti-Flag antibody immobilized magnetic beads is valuable because of its considerable reduction in nonspecific binding proteins under optimized conditions.

Name[a]	Symbol[a]	Parallel[b]	One-by-one[b]
Actin, alpha 1, skeletal muscle | Actin, alpha 2, smooth muscle, aorta | Actin, beta | Actin, alpha, cardiac muscle 1 | Actin, gamma 1 | Actin, gamma 2, smooth muscle, enteric	ACTA1 | ACTA2 | ACTB | ACTC1 | ACTG1 | ACTG2	2	2
Actin, alpha 1, skeletal muscle | Actin, alpha 2, smooth muscle, aorta | Actin, beta | Actin, gamma 2, smooth muscle, enteric	ACTA1 | ACTA2 | ACTC1 | ACTG2	1	1
ATPase family AAA domain-containing protein 3A | ATPase family AAA domain-containing protein 3B	ATAD3A | ATAD3B	6	ND
Complement component 1, q subcomponent binding protein	C1QBP	2	ND

Name[a]	Symbol[a]	Parallel[b]	One-by-one[b]
DEAH (Asp-Glu-Ala-His) box polypeptide 9	DHX9	4	ND
Eukaryotic translation elongation factor 1 alpha 1 | Eukaryotic translation elongation factor 1 alpha 2 | Eukaryotic translation elongation factor 1 alpha 1 pseudogene 5	EEF1A1 | EEF1A2 | EEF1A1P5	5	ND
Eukaryotic translation initiation factor 4A1	EIF4A1	6	2
Eukaryotic translation initiation factor 4A1 | Eukaryotic translation initiation factor 4A2	EIF4A1 | EIF4A2	2	ND
Histone cluster 1, H1c | Histone cluster 1, H1d | Histone cluster 1, H1e	HIST1H1C | HIST1H1D | HIST1H1E	2	ND
Heat shock protein 90kDa alpha (cytosolic), class A member 1	HSP90AA1	7	2
Heat shock protein 90kDa alpha (cytosolic), class A member 1 | Heat shock protein 90kDa alpha (cytosolic), class A member 2	HSP90AA1 | HSP90AA2	3	1
Heat shock protein 90kDa alpha (cytosolic), class A member 1 | Heat shock protein 90kDa alpha (cytosolic), class A member 2 | Heat shock protein 90kDa alpha (cytosolic), class B member 1 | heat shock protein 90kDa alpha (cytosolic), class B member 2 (pseudogene)	HSP90AA1 | HSP90AA2 | HSP90AB1 | HSP90AB2P	2	2
Heat shock protein 90kDa alpha (cytosolic), class A member 1 | Heat shock protein 90kDa alpha (cytosolic), class B member 1	HSP90AA1 | HSP90AB1	2	2
Heat shock protein 90kDa alpha (cytosolic), class B member 1	HSP90AB1	4	1
Heat shock protein 90kDa alpha (cytosolic), class B member 1 | Heat shock protein 90kDa alpha (cytosolic), class B member 3 (pseudogene)	HSP90AB1 | HSP90AB3P	3	1
Heat shock 70kDa protein 1A | Heat shock 70kDa protein 1B	HSPA1A | HSPA1B	13	5
Heat shock 70kDa protein 1A | Heat shock 70kDa protein 1B | Heat shock 70kDa protein 1-like	HSPA1A | HSPA1B | HSPA1L	4	3

Name[a]	Symbol[a]	Parallel[b]	One-by-one[b]
Heat shock 70kDa protein 5 (glucose-regulated protein, 78kDa)	HSPA5	19	8
Heat shock 70kDa protein 8	HSPA8	12	9
Heat shock 60kDa protein 1 (chaperonin)	HSPD1	21	10
Nucleolin	NCL	13	2
Nucleophosmin (nucleolar phosphoprotein B23, numatrin)	NPM1	3	1
Poly(A) binding protein, cytoplasmic 1	PABPC1	3	ND
Poly (ADP-ribose) polymerase 1	PARP1	15	4
Ribosomal protein L10a	RPL10A	3	ND
Ribosomal protein L11	RPL11	4	ND
Ribosomal protein L12	RPL12	4	ND
Ribosomal protein L13	RPL13	2	ND
Ribosomal protein L18	RPL18	3	1
Ribosomal protein L22	RPL22	2	1
Ribosomal protein L23	RPL23	2	1
Ribosomal protein L23a	RPL23A	5	ND
Ribosomal protein L24	RPL24	2	1
Ribosomal protein L28	RPL28	2	1
Ribosomal protein L29	RPL29	4	ND
Ribosomal protein L3	RPL3	5	ND
Ribosomal protein L30	RPL30	2	ND
Ribosomal protein L31	RPL31	3	ND
Ribosomal protein L35	RPL35	2	ND
Ribosomal protein L37a	RPL37A	2	ND
Ribosomal protein L38	RPL38	3	ND
Ribosomal protein L4	RPL4	5	2
Ribosomal protein L5 | Ribosomal protein, large, P0	RPL5 | RPLP0	5	2
Ribosomal protein L6	RPL6	6	ND
Ribosomal protein L7a	RPL7A	3	2
Ribosomal protein L8	RPL8	2	ND
Ribosomal protein L9	RPL9	3	1
Ribosomal protein, large, P0	RPLP0	2	ND
Ribosomal protein, large, P2	RPLP2	4	2
Ribosomal protein S11	RPS11	2	ND
Ribosomal protein S12	RPS12	2	ND
Ribosomal protein S13	RPS13	5	1
Ribosomal protein S15	RPS15	2	ND
Ribosomal protein S16	RPS16	4	ND
Ribosomal protein S19	RPS19	4	2
Ribosomal protein S20	RPS20	3	ND
Ribosomal protein S23	RPS23	2	ND

Name[a]	Symbol[a]	Parallel[b]	One-by-one[b]
Ribosomal protein S24	RPS24	2	ND
Ribosomal protein S25	RPS25	4	ND
Ribosomal protein S27a | Ubiquitin A-52 residue ribosomal protein fusion product 1 | Ubiquitin B | Ubiquitin C	RPS27A | UBA52 | UBB | UBC	3	2
Ribosomal protein S3	RPS3	5	2
Ribosomal protein S3A	RPS3A	7	2
Ribosomal protein S4, X-linked	RPS4X	4	1
Ribosomal protein S4, X-linked | Ribosomal protein S4, Y-linked 1 | Ribosomal protein S4, Y-linked 2	RPS4X | RPS4Y1 | RPS4Y2	3	ND
Ribosomal protein S5	RPS5	3	ND
Ribosomal protein S6	RPS6	4	ND
Ribosomal protein S7	RPS7	4	2
Ribosomal protein S8	RPS8	3	ND
Ribosomal protein S9	RPS9	4	1
Ubiquitin A-52 residue ribosomal protein fusion product 1	UBA52	3	1

Table 2. Comparison of nonspecific proteins identified from samples prepared by parallel and one-by-one methods (analyzed by MS). [a]The nonspecific proteins co-purified with β-catenin (n = 10) and Axin 1 (n = 10) using each method were categorized according to the criteria reported by Chen and Gingras. Protein Symbols and Names refer to the NCBI Gene database. Proteins identified by a common peptide sequence are indicated by ' | ' in the Name, Symbol columns. [b]Total number of the identified peptides. ND: not detected.

5. Conclusion

We have described a one-by-one sample preparation method for MS-based high-precision protein network analysis. To perform a pilot feasibility study of the one-by-one method, we designed and developed a fully automated robotic system. This system makes it possible to prepare samples under equally fast and gentle conditions. To clarify the importance of the one-by-one method, we compared protein complexes prepared by the automated one-by-one system with manual parallel preparation using β-catenin and Axin1 as baits, which are well-characterized Wnt signaling pathway proteins. One-by-one purification resulted in a sharp decrease in proteolytic degradation of purified proteins and in nonspecific binding proteins, allowing the reproducible identification of known interaction partners, as well as novel component proteins. These results suggest that one-by-one sample preparation by the automated system is useful for obtaining reliable data for high-precision analysis of protein identification and quantification for large-scale protein network analysis compared with manual parallel preparation.

We expect that this system will allow highly sensitive analyses of protein interactions using various types of cells, such as embryonic stem (ES), neuronal, and primary cells, which are limited in supply. Furthermore, we envision that this system could be used for qualitative and quantitative protein interaction network studies including chemical proteomics (Rix & Superti-Furga, 2009).

In future work, we will develop a multi-purpose robotic system that can be flexibly customized. Finally, our goal is to develop an automated robotic system that can operate not only in affinity purification, but also in general proteomics.

6. Acknowledgment

We thank H. Shibuya (Medical Research Institute, Tokyo Medical and Dental University) for providing the Flag-tagged human β-catenin and Axin1 cDNAs, and Y. Hioki, K. Koike, K. Nishimura, T. Asano and H. Kusano for technical assistance. This work was supported by a 'Development of Basic Technology to Control Biological Systems Using Chemical Compounds' grant from the New Energy and Industrial Technology Development Organization (NEDO), Japan.

7. References

Aebersold, R. & Mann, M. (2003). Mass spectrometry-based proteomics. *Nature*, 422 (6928), 198-207.

Alterovitz, G., Liu, J., Chow, J. & Ramoni, M. F. (2006). Automation, parallelism, and robotics for proteomics. *Proteomics*, 6 (14), 4016-4022.

Blagoev, B., Kratchmarova, I., Ong, S. E., Nielsen, M., Foster, L. J. & Mann, M. (2003). A proteomics strategy to elucidate functional protein-protein interactions applied to EGF signaling. *Nat. Biotechnol.*, 21 (3), 315-318.

Blow, N. (2008). Lab automation: tales along the road to automation. *Nat. Methods*, 5 (1), 109-112.

Burckstummer, T., Bennett, K. L., Preradovic, A., Schutze, G., Hantschel, O., Superti-Furga, G. & Bauch, A. (2006). An efficient tandem affinity purification procedure for interaction proteomics in mammalian cells. *Nat. Methods*, 3 (12), 1013-1019.

Chen, G. I. & Gingras, A. C. (2007). Affinity-purification mass spectrometry (AP-MS) of serine/threonine phosphatases. *Methods*, 42 (3), 298-305.

Daugherty, R. L. & Gottardi, C. J. (2007). Phospho-regulation of Beta-catenin adhesion and signaling functions. *Physiology (Bethesda)*, 22, 303-309.

Domon, B. & Aebersold, R. (2006). Mass spectrometry and protein analysis. *Science*, 312 (5771), 212-217.

Einhauer, A. & Jungbauer, A. (2001). The FLAG peptide, a versatile fusion tag for the purification of recombinant proteins. *J. Biochem. Biophys. Methods*, 49 (1-3), 455-465.

Ewing, R. M., Chu, P., Elisma, F., Li, H., Taylor, P., Climie, S., McBroom-Cerajewski, L., Robinson, M. D., O'Connor, L., Li, M., Taylor, R., Dharsee, M., Ho, Y., Heilbut, A., Moore, L., Zhang, S., Ornatsky, O., Bukhman, Y. V., Ethier, M., Sheng, Y., Vasilescu, J., Abu-Farha, M., Lambert, J. P., Duewel, H. S., Stewart, I. I., Kuehl, B., Hogue, K., Colwill, K., Gladwish, K., Muskat, B., Kinach, R., Adams, S. L., Moran, M. F., Morin, G. B., Topaloglou, T. & Figeys, D. (2007). Large-scale mapping of human protein-protein interactions by mass spectrometry. *Mol. Syst. Biol.*, 3, 89.

Gavin, A. C., Bosche, M., Krause, R., Grandi, P., Marzioch, M., Bauer, A., Schultz, J., Rick, J. M., Michon, A. M., Cruciat, C. M., Remor, M., Hofert, C., Schelder, M., Brajenovic, M., Ruffner, H., Merino, A., Klein, K., Hudak, M., Dickson, D., Rudi, T., Gnau, V., Bauch, A., Bastuck, S., Huhse, B., Leutwein, C., Heurtier, M. A., Copley, R. R., Edelmann, A., Querfurth, E., Rybin, V., Drewes, G., Raida, M., Bouwmeester, T., Bork, P., Seraphin, B., Kuster, B., Neubauer, G. & Superti-Furga, G. (2002). Functional organization of the yeast proteome by systematic analysis of protein complexes. *Nature*, 415 (6868), 141-147.

Hirano, Y., Hendil, K. B., Yashiroda, H., Iemura, S., Nagane, R., Hioki, Y., Natsume, T., Tanaka, K. & Murata, S. (2005). A heterodimeric complex that promotes the assembly of mammalian 20S proteasomes. *Nature*, 437 (7063), 1381-1385.

Ho, Y., Gruhler, A., Heilbut, A., Bader, G. D., Moore, L., Adams, S. L., Millar, A., Taylor, P., Bennett, K., Boutilier, K., Yang, L., Wolting, C., Donaldson, I., Schandorff, S., Shewnarane, J., Vo, M., Taggart, J., Goudreault, M., Muskat, B., Alfarano, C., Dewar, D., Lin, Z., Michalickova, K., Willems, A. R., Sassi, H., Nielsen, P. A., Rasmussen, K. J., Andersen, J. R., Johansen, L. E., Hansen, L. H., Jespersen, H., Podtelejnikov, A., Nielsen, E., Crawford, J., Poulsen, V., Sorensen, B. D., Matthiesen, J., Hendrickson, R. C., Gleeson, F., Pawson, T., Moran, M. F., Durocher, D., Mann, M., Hogue, C. W., Figeys, D. & Tyers, M. (2002). Systematic identification of protein complexes in Saccharomyces cerevisiae by mass spectrometry. *Nature*, 415 (6868), 180-183.

Huang, H. & He, X. (2008). Wnt/beta-catenin signaling: new (and old) players and new insights. *Curr. Opin. Cell Biol.*, 20 (2), 119-125.

Huang, S. M., Mishina, Y. M., Liu, S., Cheung, A., Stegmeier, F., Michaud, G. A., Charlat, O., Wiellette, E., Zhang, Y., Wiessner, S., Hild, M., Shi, X., Wilson, C. J., Mickanin, C., Myer, V., Fazal, A., Tomlinson, R., Serluca, F., Shao, W., Cheng, H., Shultz, M., Rau, C., Schirle, M., Schlegl, J., Ghidelli, S., Fawell, S., Lu, C., Curtis, D., Kirschner, M. W., Lengauer, C., Finan, P. M., Tallarico, J. A., Bouwmeester, T., Porter, J. A., Bauer, A. & Cong, F. (2009). Tankyrase inhibition stabilizes axin and antagonizes Wnt signalling. *Nature*, 461 (7264), 614-620.

Iioka, H., Iemura, S., Natsume, T. & Kinoshita, N. (2007). Wnt signalling regulates paxillin ubiquitination essential for mesodermal cell motility. *Nat. Cell Biol.*, 9 (7), 813-821.

Kaneko, T., Hamazaki, J., Iemura, S., Sasaki, K., Furuyama, K., Natsume, T., Tanaka, K. & Murata, S. (2009). Assembly pathway of the Mammalian proteasome base subcomplex is mediated by multiple specific chaperones. *Cell*, 137 (5), 914-925.

Kitajima, T. S., Sakuno, T., Ishiguro, K., Iemura, S., Natsume, T., Kawashima, S. A. & Watanabe, Y. (2006). Shugoshin collaborates with protein phosphatase 2A to protect cohesin. *Nature*, 441 (7089), 46-52.

Kocher, T. & Superti-Furga, G. (2007). Mass spectrometry-based functional proteomics: from molecular machines to protein networks. *Nat. Methods*, 4 (10), 807-815.

Komatsu, M., Waguri, S., Koike, M., Sou, Y. S., Ueno, T., Hara, T., Mizushima, N., Iwata, J. I., Ezaki, J., Murata, S., Hamazaki, J., Nishito, Y., Iemura, S., Natsume, T., Yanagawa, T., Uwayama, J., Warabi, E., Yoshida, H., Ishii, T., Kobayashi, A., Yamamoto, M., Yue, Z., Uchiyama, Y., Kominami, E. & Tanaka, K. (2007). Homeostatic levels of p62 control cytoplasmic inclusion body formation in autophagy-deficient mice. *Cell*, 131 (6), 1149-1163.

Komatsu, M., Kurokawa, H., Waguri, S., Taguchi, K., Kobayashi, A., Ichimura, Y., Sou, Y. S., Ueno, I., Sakamoto, A., Tong, K. I., Kim, M., Nishito, Y., Iemura, S., Natsume, T., Ueno, T., Kominami, E., Motohashi, H., Tanaka, K. & Yamamoto, M. (2010). The selective autophagy substrate p62 activates the stress responsive transcription factor Nrf2 through inactivation of Keap1. *Nat. Cell Biol.*, 12 (3), 213-223.

Natsume, T., Yamauchi, Y., Nakayama, H., Shinkawa, T., Yanagida, M., Takahashi, N. & Isobe, T. (2002). A direct nanoflow liquid chromatography-tandem mass spectrometry system for interaction proteomics. *Anal. Chem.*, 74 (18), 4725-4733.

Nishiyama, M., Oshikawa, K., Tsukada, Y., Nakagawa, T., Iemura, S., Natsume, T., Fan, Y., Kikuchi, A., Skoultchi, A. I. & Nakayama, K. I. (2009). CHD8 suppresses p53-mediated apoptosis through histone H1 recruitment during early embryogenesis. *Nat. Cell Biol.*, 11 (2), 172-182.

Olsen, J. V. & Mann, M. (2011). Effective representation and storage of mass spectrometry-based proteomic data sets for the scientific community. *Sci. Signal.*, 4 (160), pe7.

Rao, T. P. & Kuhl, M. (2010). An updated overview on Wnt signaling pathways: a prelude for more. *Circ. Res.*, 106 (12), 1798-1806.

Rix, U. & Superti-Furga, G. (2009). Target profiling of small molecules by chemical proteomics. *Nat. Chem. Biol.*, 5 (9), 616-624.

Selbach, M. & Mann, M. (2006). Protein interaction screening by quantitative immunoprecipitation combined with knockdown (QUICK). *Nat. Methods*, 3 (12), 981-983.

3

Approaches to Analyze Protein-Protein Interactions of Membrane Proteins

Sabine Hunke* and Volker S. Müller

Molekulare Mikrobiologie, Universität Osnabrück, Osnabrück, Germany

1. Introduction

About one quarter of an organismal genome encodes membrane proteins that play key roles in signal transduction, transport, energy recruitment and virulence traits of bacterial pathogens (Jones 1998; Krogh et al. 2001). The significance of membrane proteins is reflected by the fact that about 60% of all pharmaceuticals target membrane proteins (Bakheet and Doig 2009; Yildirim et al. 2007).

It can be estimated that most membrane proteins function in complexes (Fig. 1) (Daley 2008). Protein-protein interactions (PPIs) within these complexes can either be direct (primary interaction) or indirect (secondary interaction). Direct interactions occur either by homo-oligomerisation as determined for bacterial two-component systems (Gao and Stock 2009) (Fig. 1A) or by hetero-oligomerisation as shown for transport systems like ATP-binding cassette (ABC) transporters (Figs. 1B and 1C). Indirect interactions exist in large complexes as exemplified in energy producing systems such as the photosystem, bacterial surface appendages such as flagella, or secretion systems that even span two membrane systems in Gram-negative bacteria (Fig. 1D) (Jordan et al. 2001; Erhardt, Namba, and Hughes 2010). These high affinity, stable PPIs are important to form stable functional complexes (Jura et al. 2011). In addition, low affinity, transient PPIs are needed for proteins that regulate the activity of a stable complex and have been described for the interaction between e.g. ABC protein and inhibitory EIIa^{Glc} (Blüschke et al. 2007; Blüschke, Volkmer-Engert, and Schneider 2006), substrate binding protein and ABC transporter (Locher, Lee, and Rees 2002) or accessory proteins in two-component systems (Heermann and Jung 2010; Buelow and Raivio 2010; Zhou et al. 2011).

Thus, there is high demand for techniques to screen for interactions partners of and to characterize the interaction with a specific membrane protein. However, due to the hydrophobic nature of membrane proteins application of classical approaches is far more challenging than for soluble proteins (Daley 2008). Recent reviews summarize and discuss approaches to investigate PPIs for soluble proteins (Lalonde et al. 2008; Miernyk and Thelen 2008). Here, we give an overview on the current techniques used to determine and characterize PPIs of membrane proteins.

---

* Corresponding Author

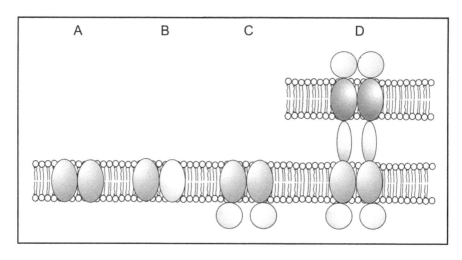

Fig. 1. Protein-protein interaction (PPIs) of membrane proteins.
Membrane proteins can be assembled as homo (A) or hetero (B) oligomers. (C) In addition,
PPIs of membrane proteins exist with peripheral proteins on either site of the membrane.
(D) Some membrane proteins are part of huge multi-protein complexes that can even span
two membranes such as secretion systems in Gram-negative bacteria (Filloux 2011).

2. Determination of membrane protein-protein interactions

Several genetic and biochemical techniques have been developed to determine PPIs of
membrane proteins. In general, these approaches use either protein fragment
complementation assays (PCA) or combine affinity purification with mass spectrometry
analysis (AP-MS). PCAs are based on the reconstituted interaction of a protein function
(reporter) by fusing two proteins of interest to complementary fragments of the reporter
protein (Ladant and Karimova 2000; Remy and Michnick 2007). AP-MS analysis allows the
determination of indirect interactions in a complex. A major advantage of these techniques
is that the protocols can be applied to almost any cell type or organism. Noteworthy, all AP-
MS approaches need a rather large amount of material and a suitable affinity tag. Both types
of screening methods, PCA and AP-MS, enable high-throughput analysis. However, the use
of high-throughput approaches may compass high levels of false-positive results and
consequently, novel PPI partners identified have to be confirmed by alternative methods
(Lalonde et al. 2008; Miernyk and Thelen 2008).

2.1 Genetic systems to analyze membrane protein-protein interactions in eukaryotes

Genetic systems established to investigate PPIs in eukaryotes use PCA. Based on classic
yeast two-hybrid systems (Fields and Song 1989) that are limited because the fusion proteins
have to be translocated into the nucleus to activate a reporter gene, new adequate methods
have been developed to overcome this limitation and allow now the analysis of membrane
protein PPIs in eukaryotes.

2.1.1 Protein-fragment complementation assays

Johnsson and Varsharvsky invented the split-ubiquitin yeast two-hybrid system (SU-YTH), a system using the endogenous mechanism of cleavage of ubiquitin by ubiquitin-specific proteases (UBPs) (Johnsson and Varshavsky 1994; Johnsson and Varshavsky 1994). Ubiquitin is the recognition marker for UBPs and can be separated into the C-terminal (Cub) and the N-terminal (Nub) part when both prey and bait are in close proximity, These two parts fused to a bait and prey protein are able to reassociate spontaneously to a quasi-native ubiquitin-molecule which can be recognized by UBPs. These proteases cleave the C-terminal attached reporter polypeptide from Cub and thereby enable the reporter transcription factor to translocate into the nucleus and to activate the reporter genes.

A second PCA represents the dihydrofolate reductase (DHFR) strategy which is also called survival selection strategy (Ear and Michnick 2009). Bait and prey proteins are fused to corresponding fragments of a modified DHFR, insensitive to methotrexate which is reconstituted and active if both interaction partners are in close proximity. The proliferation of the cells is depend on DHFR which catalyzes the reduction of dihydrofolate to tetrahydropholate during the synthesis of nucleotides and several amino acids and can be inhibited by methotexate (Remy, Campbell-Valois, and Michnick 2007; Pelletier, Campbell-Valois, and Michnick 1998). Thus only cell carrying interacting fragments of the mutated DHFR which are reassembled due to the interaction of the bait and prey proteins are able to proliferate and survive in the presence of the inhibitor methotrexate.

Besides the DHFR, different reporter enzymes can be used for a PCA strategy e.g. yeast cyosine deaminase (OyCD) (Ear and Michnick 2009) or fluorescent proteins (see chapter 3.6.3) extensively reviewed (Michnick et al. 2011).

2.1.2 Reverse Ras recruitment system (reverse RRS)

An alternative method to SU-YTH is the reverse Ras recruitment system (reverse RRS) (Hubsman, Yudkovsky, and Aronheim 2001) that is based on Ras recruitment system (RRS) in yeast (Broder, Katz, and Aronheim 1998). Growth of yeast depends on cAMP. cAMP is generated by adenylate cyclase which is activated by Ras which itself is activated by Cdc25 (Cannon, Gibbs, and Tatchell 1986). In contrast to cytoplasmic Ras, membrane-bound Ras complement a temperature-sensitive mutant in Cdc25 (Aronheim et al. 1997; Aronheim 1997) (Petitjean, Hilger, and Tatchell 1990). PPI between a membrane protein and its interaction partner results in Ras translocation and allows cell growth at elevated temperature (Aronheim 2001). Thus, reverse RRS can be used to screen for a soluble protein as PPI partner for a membrane protein

2.2 Genetic systems to analyze membrane protein-protein interactions in bacteria

PCA can also be used as genetic systems to analyze membrane protein-protein interactions in bacteria. PCAs based on a protein function directly involved in transcription are restricted to determine the interaction of soluble proteins (bacteriophage lambda repressor λcI, E. coli LexA repressor, DNA loop formation, RNA polymerase recruitment) (Ladant and Karimova 2000). In contrast, PCAs based on metabolism or signaling cascades can be adapted for membrane proteins (bacterial mDHFR survival assay, BACTH).

2.2.1 Murine dihydrofolate reductase (mDHFR)

A PCA based on the essential DHFR has also been established to screen for PPI in bacteria. Prokaryotic DHFR but not murine DHFR (mDHFR) is inhibited by trimethoprim (Appleman et al. 1988). PPI of the two proteins of interest fused to mDHFR fragments allow *E. coli* to grow on media supplemented with trimethoprim (Remy, Campbell-Valois, and Michnick 2007), allowing a positive selection.

2.2.2 Bacterial adenylate cyclase two-hybrid assay (BACTH)

The bacterial adenylate cyclase two-hybrid assay (BACTH) is well established to investigate PPIs for membrane proteins in bacteria (Fig. 2) (Karimova, Dautin, and Ladant 2005). BACTH is based on the reconstituted interaction of two *Bordetella pertussis* adenylate cyclase fragments (T18 and T25) resulting in elevated levels of cAMP (Karimova et al. 1998). cAMP is a key signaling molecule in *E. coli* that activates the catabolite activator protein (CAP) resulting in transcriptional activation of metabolic operons such as those for lactose and maltose (Deutscher, Francke, and Postma 2006). Consequently, PPI of two membrane proteins fused to adenylate cyclase fragments results in fermentation of lactose or maltose which can easily be detected on either indicator (MacConkey maltose or X-Gal plates) or selection media (minimal media supplemented with either lactose or maltose as carbon source) (Karimova, Dautin, and Ladant 2005). In addition, BACTH allows quantification of the PPI by measuring the activity of the lactose cleaving β-galactosidase (Robichon et al. 2011).

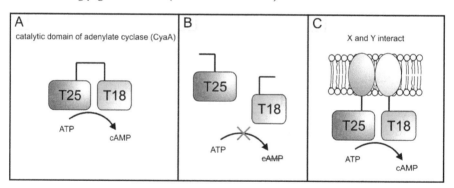

Fig. 2. The bacterial adenylate cyclase two-hybrid assay (BACTH). (A) The adenylate cyclase (CyaA) of *Bordetella pertussis* synthesizes cAMP in *E. coli*. cAMP activates the catabolite activator protein resulting in target gene expression. (B) Coexpression of two CyaA fragments (T25, T18) does not result in protein fragment complementation (PCA). (C) Fusion of T25 and T18 to interacting membrane proteins results in PCA and cAMP production (Karimova et al. 1998; Karimova, Dautin, and Ladant 2005).

2.3 Co-immunoprecipitation

Co-immunoprecipitation (co-IP) is the classical method to screen for and the proof of PPIs. On the one hand, co-IP is a highly specific, yet relatively simple technique that allows the identification of two or more proteins *in vivo* (Miernyk and Thelen 2008). On the other hand, co-IP requires an antibody with high specificity for the protein of interest. However, highly specific antibodies for membrane proteins are even more difficult to obtain than for soluble proteins. Therefore, when using co-IP for membrane proteins false-positive results are reduced by pre-clearing solubilized membrane proteins by the addition of immobilized protein A or protein G (protein A/G) (Vaidyanathan et al. 2010). The pre-cleared supernatant is then incubated with the primary antibodies and protein-antibody complexes are formed. The protein-antibody complexes are recovered using immobilized protein A/G. Using co-IP mainly strong PPIs as found in complexes can be detected, but transient interactions are rather difficult to be determined.

2.4 Tandem affinity purification (TAP)

One AP-MS technique is the tandem affinity purification (TAP) method (Xu et al. 2010; Puig et al. 2001; Rigaud, Pitard, and Levy 1995). TAP allows the identification of direct interactions in a protein complex and uses the fusion of a TAP tag to the either the N or the C terminus of the target protein (Xu et al. 2010). The TAP tag consists of a calmodulin-binding domain (CBD), a cleavage site for the tobacco etch virus (TEV), and the IgG binding units of the protein A of *Staphylococcus aureus*. Protein complexes containing a TAP tagged protein can be purified by two very specific purification steps. In the first purification step the TAP tagged complex is bound to an IgG column. Elution occurs by cleaving off the protein complex from the column using the TEV cleavage site of the TAP tag. In the second purification step, the TAP-tagged complex is bound to calmodulin beads. After EGTA elution the complex can be further analyzed with respect to the interaction partners of the TAP-tagged bait protein. When using mild detergent the TAP method can also be adapted for membrane proteins. However, the TAP system is considered to be inefficient in identifying transient interactions (Xu et al. 2010). A very recent review describes the application and limits of TAP in detail (Xu et al. 2010).

2.5 Chemical cross-linking and mass spectrometry techniques

2.5.1 Protein Interaction Reporter (PIR) technology

IP-based affinity purification methods always require the genetically introduction of a tag fused to a target protein of interest. The overexpression of the fusion proteins can lead to improper intercellular localization and by this to false positives (Bouwmeester et al. 2004). Furthermore, the co-elution of potential interaction partners of a target protein (see chapter 2.3 and 2.4) is often negatively affected during the purification of the target proteins afterwards. The protein interaction reporter (PIR) technology established by Xiaoting Tang and James E. Bruce overcomes these limitations by the use of new design of crosslinker. These cross-linkers include two reactive groups to cross link potential interaction partners, two labile bonds and a mass encoded reporter containing an affinity tag. The labile bonds can be cleaved afterwards by UV irradiation prior to the identification of interaction partners. The applications of the PIR technology have been extensively reviewed (Hoopmann, Weisbrod, and Bruce 2010; Tang and Bruce 2010; Yang et al. 2010).

2.5.2 Membrane strep–protein interaction experiment (SPINE)

Membrane-SPINE is an improved technique based on the Strep-protein interaction experiment (SPINE) adapted to membrane proteins (Herzberg et al., 2007). It combines the fixation of protein complexes in a cell by formaldehyde cross-linking *in vivo* with the specific purification of a Strep-tagged target membrane protein (Müller et al. 2011). Due to its small size formaldehyde can easily penetrate membranes and create an effective snap shot of the interactome of a living cell (Fig. 3). Thus not only the target protein but also cross-linked potential interaction partner can be co-eluted (Fig. 3 B) and identified afterwards by Mass spectrometry (Fig. 3 E) or immunoblot analysis (Fig. 3 D). By using Membrane-SPINE it is possible to monitor not only permanent protein-protein interactions but also transient interactions occurring during signal transduction (Müller et al. 2011).

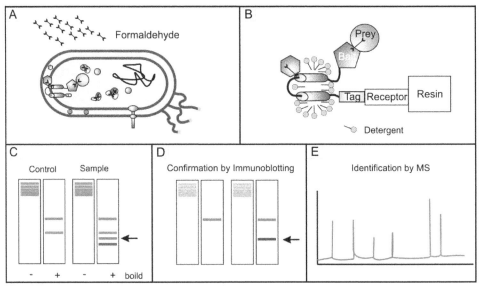

Fig. 3. Membrane-SPINE. (A) Protein complexes of a living cell are fixed by formaldehyde cross-linking. (B) After detergent solubilization the strep-tagged membrane protein is purified via a strep-tactin resin. (C) Co-eluted interaction partners of the strep-tagged membrane protein can be separated after boiling through SDS-PAGE and finally confirmed by immunoblotting (D) or identified by MS analysis (E) (Müller et al. 2011).

2.6 *In silico* prediction of membrane protein-protein interactions

The recent progress in the fields of bioinformatics has culminated in the development of powerful tools for the prediction of protein-protein interactions *in silico*. The growing amount of data of protein interactions and protein sequence information have been successfully used for the prediction of new protein interaction networks by the homogenous protein mapping method (Saeed and Deane 2008) or co-evolution analysis (Skerker et al. 2008). By using a computational approach Skerker and colleagues compared by sequence alignment nearly 1300 pairs of histidine kinases (HK) and response regulators (RR) of two-component systems of almost 200 sequenced bacterial genomes to identify the structural basis determining the interaction between HK and RR (Skerker et al. 2008). In the same line, Procaccini and coworkers identified a molecular interaction code between HK and RR by comparing 8998 paired SK/RR sequences were of 769 fully sequenced bacterial genomes which results in a highly specific preference between both interaction partners. Based on this code, it was possible to predict clusters of cross-talk candidates between non-cognate signaling partners (Procaccini et al., 2011). However, all these predictions can strongly contradict the situation *in vivo* which emphasizes the importance of the methods presented within this review for verification.

3. Characterization of membrane protein-protein interactions

Different methods are available to characterize PPIs between membrane proteins. They allow the investigation of functional interaction, kinetics, and affinities between membrane proteins. Moreover, the dynamics of the interaction, the interface between the proteins and the cellular localization can be determined.

3.1 Reconstitution of membrane proteins for functional studies

For detailed biochemical investigation of membrane proteins the incorporation of the purified proteins into a lipid bilayer is essential. This technique is also known as reconstitution and results in proteoliposomes that allow the characterization of membrane proteins without the influence of other membrane components (Rigaud 2002; Rigaud and Levy 2003; Paternostre, Roux, and Rigaud 1988). The importance of reconstitution results from the observation that many membrane proteins are only fully active when incorporated into a lipid bilayer (Rigaud 2002; Fleischer et al. 2007).

Basis for any successful reconstitution are the quality of the purified membrane protein, the lipid requirement of the membrane protein and the ratio between lipid and protein (Geertsma et al. 2008; Knol, Sjollema, and Poolman 1998). The usage of mild detergents, such as n-dodecyl-β-maltoside (DDM) or Triton X-100, is highly recommended to reduce dissociation during purification and to keep by this the complex active (Geertsma et al. 2008). The addition of chemical chaperones (glycerol, salt), phospholipids or ligands can further stabilize membrane protein complexes during the purification procedure (Geertsma et al. 2008).

To reconstitute a protein function into proteoliposomes, detergent-solubilized and purified membrane protein is mixed with detergent-destabilized lipid vesicles. In order to generate membrane vesicles and to incorporate the membrane protein into these vesicles the detergent has to be removed. Several techniques exist (dilution, dialysis, SEC) but when using mild detergents that have in general a low critical micellar concentration (CMC) the adsorption of the detergent onto polystyrene beads is the method of choice.

After reconstitution of a membrane protein into proteoliposomes several controls have to be performed: the morphology and residual permeability of the proteoliposomes have to be proved; the incorporation efficiency and the orientation of the membrane proteins have to be determined to allow kinetic studies; and the functionality of the reconstituted membrane proteins has to be proved by activity assays. We have observed during our studies with different kind of membrane proteins that not always the highest solubilization and purification efficiency results in the most active protein (Hunke et al., 1997; Fleischer et al., 2007). Notably, when working with the same membrane protein but from different organisms we had to change the detergent (Fleischer et al. 2007; Müller et al. 2011). We and others, use proteoliposomes not only to analyze the functional interaction between a membrane protein (reconstituted sensor kinase) and a soluble partner (cognate response regulator) (Fleischer et al. 2007; Jung, Tjaden, and Altendorf 1997) but also in order to investigate the impact of different conditions on this interaction (Fleischer et al. 2007).

Together, the knowledge on the physical background of lipid-detergent systems and the mechanisms of proteoliposome formation results in a number of basic principles in membrane protein reconstitution (Rigaud and Levy 2003; Silvius 1992) that allowed the establishment of general protocols (Geertsma et al. 2008).

3.2 Native electrophoretic techniques

Blue native electrophoresis (BNE; also known as Blue Native PAGE) has been developed to purify active membrane protein complexes from mitochondria (Schägger and von Jagow 1991). Therefore, membrane proteins are solubilzed using a mild neutral detergent and insoluble proteins are removed by centrifugation. Solubilized proteins are then mixed with the anionic dye Coomassie Brilliant Blue (CBB) G-250 which binds to protein surfaces (Compton and Jones 1985). Binding of CBB G-250 results in a negative charge shift and allows a protein complex to migrate into a non-denaturating polyacrylamid gel. Additional separation methods in the second dimension (2D) as e.g. denaturating SDS-PAGE can be used to separate single proteins of a membrane protein complex in order to estimate the native mass or to identify proteins of one membrane protein complex (Wittig and Schägger 2009). Because CBB interferes with the activity of proteins and the analysis of fluorescent-labeled proteins, clear-native electrophoresis (CNE) and high-resolution CNE (hrCNE) have been established (Wittig, Karas, and Schägger 2007). CNE uses no dye and proteins migrate according to their intrinsic pI (Wittig and Schägger 2009). The two disadvantages of CNE, proteins with a pI>7 are lost and smearing of the membrane proteins over the gel, can be partially compensated when applying hrCNE that uses detergent micelles to induce the charge shift (Wittig and Schägger 2008).

Extended recent reviews on BNE, CNE and hrCNE summarize the power of these techniques for the identification and characterization of PPIs for membrane proteins and explain the protocol in detail (Wittig and Schägger 2008, 2009; Krause 2006; Miernyk and Thelen 2008).

3.3 Far-Western Blot

Far-Western Blot analysis is an *in vitro* method to proof and to identify direct PPIs between two proteins (Edmondson and Roth 2001; Wu, Li, and Chen 2007).

When using Far-Western blot analysis to verify PPI, one protein is immobilized on a continuous membrane sheet and the blot is incubated with a purified second protein, the bait protein. Afterwards, the blot is treated as a normal immunoblot using an antibody against the bait protein. When using Far-Western blot analysis to identify a PPI between a bait protein and a prey protein in a cell lysate, the cell lysate is transferred to a continuous membrane sheet instead of a purified protein. Then, the blot is incubated with the bait protein and finally treated as an immune blot against the bait protein.

Although this method is most suitable for soluble proteins it can also be adapted to membrane proteins. However, detergent interferes with the immunoblot procedure (Zhou et al. 2011). Thus, when using Far-Western blot analysis for membrane proteins either the purified membrane protein or the membrane fractions should be immobilized to the continuous membrane sheet as the prey. In other words, Far-Western blot analysis allows the identification of a PPI between a membrane protein (in a membrane fraction) and a soluble bait protein without the need of purifying a membrane protein by detergent treatment.

Detailed protocols for the Fat-Western blot procedure are given by Edmondson & Roth (2001) and Wu et al. (2007).

3.4 SPOT-analysis

SPOT-analysis is an *in situ* screening technique developed for the identification of interacting epitopes (Frank 1992, 2002; Volkmer 2009). For SPOT-analysis a peptide array is generated by coupling single amino acids step by step first on a continuous membrane sheet then on the first amino acid and so on (Frank 2002; Reimer, Reineke, and Schneider-Mergener 2002; Wenschuh et al. 2000). By this, SPOT-analysis allows the rapid and parallel synthesis of different synthetic peptides that can be analyzed simultaneously. Even more important, SPOT-analysis permits the substitutional analysis of an epitope without the need of mutagenesis and purification (Volkmer 2009). As for Far-Western blot analysis the peptide array is first incubated with the bait protein and finally developed as normal immunoblot versus the bait protein. However, this elegant technique is limited on hydrophilic peptides and cannot be used to characterize PPIs of TMSs. Never the less, different groups have used SPOT-analysis to screen for and to identify epitopes in hydrophilic domains of membrane proteins important for the interaction with soluble proteins (Zhou et al. 2011; Blüschke, Volkmer-Engert, and Schneider 2006).

Generation of peptide arrays in macro- and micro-array format is described in detail by the groups of Frank and Schneider-Mergener (Frank 2002; Reimer, Reineke, and Schneider-Mergener 2002; Wenschuh et al. 2000).

3.5 Surface Plasmon Resonance (SPR)

The most elegant approach to quantify binding kinetics, thermodynamics and concentrations in PPIs *in vitro* is the surface plasmon resonance (SPR) technology. This technique needs both proteins to be purified.

When using SPR for soluble proteins, one purified protein is bound to a gold-coated surface of a chip. To obtain the background, the chip is floated with the buffer the second protein is purified with and the refractive index of the solvent near the gold surface is measured. In the next step the second purified protein in its buffer is floated and the refractive index is again measured. PPI is determined by the changes in refractive index.

Because membrane proteins cannot be bound to chip surface as efficient as soluble proteins, specific chips have been designed that allow the capture of proteoliposomes (Maynard et al. 2009). A detailed protocol for this approach is given by (Hodnik and Anderluh 2010). Technologies that allow the analysis of membrane proteins directly on a chip are still in develop (Maynard et al. 2009). Nevertheless SPR can already be utilized to analyze PPI between a membrane protein and a soluble protein. Therefore, the soluble protein is immobilized to a classical SPR chip and floated with proteoliposomes containing the membrane protein. This experimental setup has successfully been used to characterize the transient interaction between the membrane integral sensor kinase KdpD and the scaffolding protein UspC (Heermann et al. 2009).

3.6 Imaging technologies

Imaging technologies allow the characterization of PPIs in the native environment of proteins *in vivo* and in real time. Thus, they are excellent tools to study mechanisms in protein function.

In general, imaging technologies use genetic fusions between the protein of interest and a fluorescent protein (Fig. 4). Genetic fusions are easily applicable for any protein including membrane proteins. However, for membrane proteins it has to be taken into account that fluorescent proteins like the green fluorescent protein (GFP) are only folded correctly in the cytosol. Consequently, fusions between membrane proteins and fluorescent proteins should only be performed at those domains of a membrane protein known to be localized inside the cell.

Initial experiments to analyze PPIs with imaging technologies are co-localization studies of two-labeled proteins in order to determine their cellular distribution. Subsequently, a variety of imaging technologies can be used to characterize PPIs of membrane proteins in more detail (Lalonde et al. 2008; Schäferling and Nagl 2011).

3.6.1 Fluorescence resonance energy transfer (FRET)-based techniques

Fluorescence (or Förster) resonance energy transfer (FRET) is a biophysical method detecting energy transfer from a donor fluorophor to an acceptor fluorophor (Fig. 4A) (reviewed in (Masi et al. 2010; Schäferling and Nagl 2011). The principle was first described by Theodor Förster, 1948. The basis of FRET is the correct donor-acceptor pair. The emission wavelength of the donor fluorophor has to be in the range of the excitation wavelength for the acceptor fluorophor. When the two fluorophores are in sufficient proximity (2-8 nm) excitation of the donor induces energy emission that can be absorbed by the acceptor resulting in a characteristic energy emission of the acceptor. Well established donor acceptor pairs in cell biology are the combination of the cyan fluorescent proten (CFP) with the yellow fluorescent protein (YFP), GFP with rhodamine, fluoresceinisothiocyanate and Cy3, and CFP with the fluorescein arsenical helix binder (FlAsH) (Hoffmann et al. 2005).

To analyze the distance and dynamics of membrane proteins, cell lines are co-transfected (bacteria are co-transformed) with two vectors carrying a CFP –bait protein and an YFP-prey protein fusion. The FRET signal reflects the PPI between bait and prey and is determined by fluorescence microscopy. Recently, FRET has additionally been demonstrated as a tool for high-throughput screening of PPIs in living mammalian cells (Banning et al. 2010). Therefore, FRET measurement was combined with fluorescence activated cell sorting (FACS). To do so, the human cell line 293T was co-transfected with a vector carrying a fusion between the human immunodeficiency virus (HIV) Vpu accessory protein and YFP and a second vector carrying a fusion between a cDNA library and CFP. Cells were sorted for a positive FRET signal and PPIs proofed by co-IP. However, an average of more than 50% false positives was estimated which is comparable with Y2H screens (Banning et al. 2010).

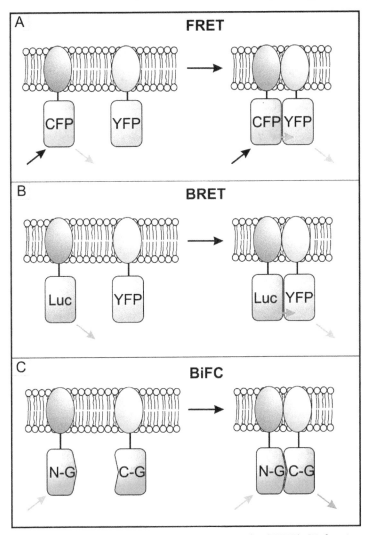

Fig. 4. Comparison of fluorescence resonance energy transfer (FRET), Bioluminescence Resonance Energy Transfer (BRET) and bimolecular fluorescence complementation (BiFC) (A) FRET: A donor fluorophore (here CFP) is fused to protein (orange) and an acceptor fluorophore (here YFP) is fused to a second protein (green). When the two proteins are in sufficient proximity fluorescence energy transfer can be monitored. (B) BRET: As in FRET, energy transfer between a donor and an acceptor is determined, but the donor is a protein that emits light (here luciferase). (C) BiFC: A fluorescent protein (here GFP) is split in two halves. Interaction of the two proteins fused to these two halves results in protein fragment complementation (PCA).

Fluorescence lifetime imaging microscopy (FLIM) is a FRET-based technique established to identify sub-cellular distributions of specific post-translational changes in protein targets (Peltan et al. 2006). In contrast to FRET, FLIM measurement determines the relaxation time of the acceptor flourophor and not the emission quantity (Biskup et al. 2007; Wouters 2006). As a consequent, FLIM measurement is independent from fluorophore concentrations and therefore the FRET-based method of choice to investigate dynamics in PPIs (Lalonde et al. 2008).

Total internal reflection fluorescence (TIRF) microscopy is a FRET-based approach used to study processes close to or at cell membranes (Mattheyses, Simon, and Rappoport 2010). In principle, TIRF results as the light beam propagates first through glass with a high refractive index and then through water with a low refractive index. As a consequence, the direction of the light beam is altered and an evanescent field is generated. Therefore, TIRF microscopy stimulates only fluorophores very close to the cover slip resulting in a minimized background fluorescence and reduced cellular photo-damage (Mattheyses, Simon, and Rappoport 2010; Lam et al. 2010).

General extended reviews on fluorescence microscopy techniques are given by Waters, North and Masi et al., (North 2006; Waters 2009; Masi et al. 2010). For FLIM background we refer to Lalonde et al. (2008). A detailed protocol and trouble-shooting for FLIM is given by Periasamy (Sun, Day, and Periasamy 2011). Detailed reviews on the physical basis of TIRF and advanced applications are given by Axelrod and Rappoport (Mattheyses, Simon, and Rappoport 2010; Axelrod 2003; Axelrod 2008).

3.6.2 Bioluminescence Resonance Energy Transfer (BRET)

Bioluminescence resonance energy transfer (BRET) is a variation of FRET using an autofluorescent protein as a donor (Fig. 4B) (Xia and Rao 2009; Pfleger and Eidne 2006). Consequently, excitation of the donor is not required. The most popular used BRET pair is a combination of coelenterazine emitting energy around 400 nm and a variant of GFP, termed GFP2 (Jensen et al. 2002).

3.6.3 Bimolecular Fluorescence Complementation (BiFC)

Bimolecular fluorescence complementation (BiFC) is fluorescence technique based on PCA (Fig. 4C). Two halves of a fluorescence protein, in general GFP (N-GFP and C-GFP), are fused to either the bait or the prey protein. PPI of bait and prey protein results in a fluorescent signal that can be monitored by fluorescence microscopy. However, BiFC cannot be used for dynamic studies because half-life time of the N-GFP and C-GFP interaction was estimated to be 10 years (Magliery et al. 2005).

3.7 Site-directed chemical cross-linking

Site-directed chemical cross-linking is a powerful tool to characterize the distance and the dynamics of specific amino acid pairs in and between membrane proteins both *in vivo* and *in vitro* (Kaback et al. 2011; Bordignon, Grote, and Schneider 2010). In many cases, homo-bifunctional sulfhydryl cross-linkers are used. These have variable spacer-arm length ranging from 5 to 50 Å. Because of their hydrophobic spacer arms, many cross-linkers are membrane-permeable and thus ideal to perform cross-linking with membrane proteins as exemplified for the maltose ABC transporter (Bordignon, Grote, and Schneider 2010). To prevent unspecific inter-molecular cross-linking, cross-linkers with a maximum spacer-arm length of 25 Å should be chosen.

Ideally, the native cysteine residues within proteins are first substituted by other amino acids (Ala or Ser) to allow specificity in site-directed chemical cross-linking. Hereafter, cysteine insertion mutagenesis is performed. The functionality of first the cysteine-free and then the mono-cysteine proteins has to be confirmed after each substitution step (Hunke et al. 2000; Hunke and Schneider 1999). Finally, the cross-linking procedure is performed either *in vivo* (Shiota et al. 2011), or *in vitro* using crude membranes or the reconstituted system (Hunke et al. 2000; Daus et al. 2007). When using the reconstituted system, substrates or inhibitors can be added during the cross-linking procedure providing information about the dynamics within a complex (Daus et al. 2007).

Comprehensive background and application for site-directed chemical cross-linking is given by the two major suppliers for cross-linking agents Pierce (http://www.piercenet.com/files/1601673_Crosslink_HB_Intl.pdf) and Molecular Probes (www.invitrogen.com/site/us/en/home/References/Molecular-Probes-The-Handbook.html).

3.8 Site-directed spin labeling electron paramagnetic resonance spectroscopy (EPR)

Site-directed spin labeling (SDSL) electron paramagnetic resonance (EPR) spectroscopy is a biophysical method introduced by Wayne L. Hubbell (Altenbach et al. 1990; Altenbach et al. 1989) that allows the determination not only of distances in and between macromolecules but also their dynamics (reviewed in Berliner et al., 2002; Klare & Steinhoff, 2010). In addition, EPR spectroscopy techniques provide a high time resolution and are independent on the protein size (reviewed in Klare & Steinhoff, 2010 and in this issue by Klare, 2012).

Spin labels are introduced at two cysteine residues in the otherwise cysteine-free complex and are excited by a strong microwave pulse. The most frequently used spin label is the methanethiosulfonate spin label (1-oxyl-2,2,5,5-tetramethyl-D3-pyrroline-3-yl)methanethiosulfonate (MTSL). The physical principle that the intensity of the dipolar interaction between the two spin labels is inversely proportional to the cube of their distance, allows the calculation of the distance between the two spin labeled residues (Fajer, Brown, and Song 2007; Klare and Steinhoff 2010; Klare 2012).

EPR spectroscopy methods (exchange EPR, dipolar continuous wave EPR) cover a distance range up to 2 nm. Moreover, dipolar continuous wave (CW) EPR spectroscopy yields information on the sidechain mobility as well as the accessibility and polarity of the microenvironment of a spin label at single labeled proteins (Bordignon and Steinhoff 2007).

Pulse dipolar EPR methods in particular double electron–electron resonance (DEER) spectroscopy allows the determination of a distance ranges from 2-8 nm in PPIs (Pannier et al. 2000). DEER uses two microwave frequencies resulting in two spin populations. Thereby, one spin population influences the echo amplitude of the second spin population

(Fajer, Brown, and Song 2007). An open-source software (DEER Analysis 2011) for extracting distance distributions from DEER data sets has been provided by the ETH-Zurich (http://www.epr.ethz.ch/software/index). DEER gave a deeper inside the transmembrane signaling mechanism of rhodopsin (Altenbach et al. 2008; Knierim et al. 2008), sensory rhodopsin (Holterhues et al. 2011; Klare et al. 2011), the maltose ABC transporter (Grote et al. 2008; Grote et al. 2009) and the KtrAB potassium transporter (Hänelt et al. 2010). Thus, during recent years and on the basis of crystallographic data, DEER has been established as the state of the art technique to allow description of signal and transport mechanisms (Bordignon, Grote, and Schneider 2010; Klare and Steinhoff 2010).

4. Conclusions and outlook

The advances in genome, proteome and *in silico* analysis have identified membrane proteins with no assigned function. Moreover, it became evident that most membrane proteins function in complexes that are composed of several subunits (Daley 2008). Elucidation of the identification and characterization of PPIs of integral membrane proteins is the challenging task of today's research. During recent years new methods have emerged that offer new opportunities to determine partner, kinetics and thermodynamics in membrane protein PPIs. Fluorescence techniques allow now the investigation of the location and interaction of membrane proteins *in vivo*. The application of EPR techniques has just started to allow a deeper inside into the mechanisms in membrane protein PPIs. Combination of the techniques presented here will allow in the future to elucidate the mechanism of signal transmission and substrate transport from one side of a membrane to the other.

5. Acknowledgment

This work was financially supported by the Deutsche Forschungsgemeinschaft. We are grateful to Michael Hensel for critical reading the manuscript.

6. References

Altenbach, C, SL Flitsch, HG Khorana, and WL. Hubbell. 1989. Structural studies on transmembrane proteins. 2. Spin labeling of bacteriorhodopsin mutants at unique cysteines. *Biochemistry* 28 (19):7806-7812.

Altenbach, Christian, Ana Karin Kusnetzow, Oliver P. Ernst, Klaus Peter Hofmann, and Wayne L. Hubbell. 2008. High-resolution distance mapping in rhodopsin reveals the pattern of helix movement due to activation. *Proc. Natl. Acad. Sci. U.S.A.* 105 (21):7439-7444.

Altenbach, Christian, T. Marti, H. Gobind Khorana, and Wayne L. Hubbell. 1990. Transmembrane Protein Structure: Spin Labeling of Bacteriorhodopsin Mutants. *Science* 248 (1088-1092).

Appleman, J R, N Prendergast, T J Delcamp, J H Freisheim, and R L Blakley. 1988. Kinetics of the formation and isomerization of methotrexate complexes of recombinant human dihydrofolate reductase. *J. Biol. Chem.* 263 (21):10304-10313.

Aronheim, A, E Zandi, H Hennemann, S.J. Elledge, and M. Karin. 1997. Isolation of an AP-1 repressor by a novel method for detecting protein-protein interactions. *Mol. Cell Biol.* 17 (6):3094-3102.

Aronheim, A. 1997. Improved efficiency sos recruitment system: expression of the mammalian GAP reduces isolation of Ras GTPase false positives. *Nucleic. Acids Res.* 25 (16):3373-3374.

Aronheim, Ami. 2001. Protein recruitment systems for the analysis of protein-protein interactions. *Methods* 24 (1):29-34.

Axelrod, D. 2003. Total internal reflection fluorescence microscopy in cell biology. *Biophotonics B* 361:1-33.

Axelrod, Daniel. 2008. Chapter 7: Total Internal Reflection Fluorescence Microscopy. In *Methods in Cell Biology*, edited by J. C. Dr. John and Dr. H. William Detrich, III: Academic Press.

Bakheet, Tala M., and Andrew J. Doig. 2009. Properties and identification of human protein drug targets. *Bioinformatics* 25 (4):451-457.

Banning, Carina, Jörg Votteler, Dirk Hoffmann, Herwig Koppensteiner, Martin Warmer, Rudolph Reimer, Frank Kirchhoff, Ulrich Schubert, Joachim Hauber, and Michael Schindler. 2010. A Flow Cytometry-Based FRET Assay to Identify and Analyse Protein-Protein Interactions in Living Cells. *PLoS ONE* 5 (2):e9344.

Biskup, C., T. Zimmer, L. Kelbauskas, and B. Hoffmann. 2007. Multi-dimensional fluorescence lifetime and FRET measurements. *Microsc. Res. Tech.* 70:442–451.

Blüschke, Bettina, Viola Eckey, Britta Kunert, Susanne Berendt, Heidi Landmesser, Michael Portwich, Rudolf Volkmer, and Erwin Schneider. 2007. Mapping putative contact sites between subunits in a bacterial ATP-binding cassette (ABC) transporter by synthetic peptide libraries. *J. Mol. Biol.* 369 (2):386-399.

Blüschke, Bettina, Rudolf Volkmer-Engert, and Erwin Schneider. 2006. Topography of the surface of the signal-transducing protein EIIAGlc that interacts with the MalK subunits of the maltose ATP-binding cassette transporter (MalFGK2) of *Salmonella typhimurium*. *J. Biol. Chem.* 281 (18):12833-12840.

Bordignon, Enrica, Mathias Grote, and Erwin Schneider. 2010. The maltose ATP-binding cassette transporter in the 21st century – towards a structural dynamic perspective on its mode of action. *Mol. Microbiol.* 77 (6):1354-1366.

Bordignon, Enrica, and H.J. Steinhoff. 2007. Membrane Protein Structure and Dynamics Studied by Site-directed Spin Labeling ESR. In *Biological Magnetic Resonance*, edited by M. A. Hemmingo and L. J. Berliner. New York: Springer.

Bouwmeester, T, A Bauch, H Ruffner, PO Angrand, G Bergamini, K Croughton, C Cruciat, D Eberhard, J Gagneur, S Ghidelli, C Hopf, B Huhse, R Mangano, AM Michon, M Schirle, J Schlegl, M Schwab, MA Stein, A Bauer, G Casari, G Drewes, AC Gavin, DB Jackson, G Joberty, G Neubauer, J Rick, B Kuster, and G. Superti-Furga. 2004. A physical and functional map of the human TNF-alpha/NF-kappa B signal transduction pathway. *Nat. Cell Biol.* 6 (2):97-105.

Broder, Yehoshua C., Sigal Katz, and Ami Aronheim. 1998. The Ras recruitment system, a novel approach to the study of protein-protein interactions. *Curr. Biol.* 8 (20):1121-1130.

Buelow, Daelynn R. , and Tracy L. Raivio. 2010. Three (and more) component regulatory systems & auxiliary regulators of bacterial histidine kinases. *Mol. Microbiol.* 75 (3):547-566.

Cannon, J.F., J.B. Gibbs, and K Tatchell. 1986. Suppressors of the ras2 mutation of Saccharomyces cerevisiae. *Genetics* 113 (2):247-264.

Compton, S.J., and C.G. Jones. 1985. Mechanism of dye response and interference in the Bradford protein assay. *Anal. Biochem.* 151 (2):369-374.

Daley, Daniel O. 2008. The assembly of membrane proteins into complexes. *Curr. Opin. Struct. Biol.* 18 (4):420-424.

Daus, M.L., M. Grote, P. Müller, M. Doebber, S. Herrmann, H.J. Steinhoff, E. Dassa, and Schneider E. 2007. ATP-driven MalK dimer closure and reopening and conformational changes of the "EAA" motifs are crucial for function of the maltose ATP-binding cassette transporter (MalFGK2). *. J Biol Chem* 282 (22387-22396).

Deutscher, Josef, Christof Francke, and Pieter W. Postma. 2006. How phosphotransferase system-related protein phosphorylation regulates carbohydrate metabolism in bacteria. *Microbiol. Mol. Biol. Rev.* 70 (4):939-1031.

Ear, Po Hien, and Stephen W. Michnick. 2009. A general life-death selection strategy for dissecting protein functions. *Nat Meth* 6 (11):813-816.

Edmondson, Diane G., and Sharon Y. Roth. 2001. Identification of protein interactions by far western analysis. In *Curr. Protoco. Mol. Biol.*: John Wiley & Sons, Inc.

Erhardt, Marc, Keiichi Namba, and Kelly T. Hughes. 2010. Bacterial Nanomachines: The Flagellum and Type III Injectisome. *Cold Spring Harbor Perspectives in Biology* 2 (11).

Fajer, P.G., L. Brown, and L. Song. 2007. Practical pulsed dipolar ESR (DEER). *in: Hemminga M.A. and Berliner L.J. (2007) ESR spectroscopy in membrane biophysics. Series: Biological Magnetic Resonance, Springer: New York* 27:95-128.

Fields, Stanley, and Ok-kyu Song. 1989. A novel genetic system to detect protein–protein interactions. *Nature* 340 (6230):245-246.

Filloux, Alain. 2011. Protein secretion systems in *Pseudomonas aeruginosa*: an essay on diversity, evolution and function. *Front. Microbiol.* 2.

Fleischer, Rebecca, Ralf Heermann, Kirsten Jung, and Sabine Hunke. 2007. Purification, reconstitution, and characterization of the CpxRAP envelope stress system of *Escherichia coli*. *J. Biol. Chem.* 282 (12):8583-8593.

Frank, R. 1992. SPOT-synthesis: an easy technique for the positionally addressable, parallel chemical synthesis on a membrane support. *Tetrahedron* 48:9217-9232.

Repeated Author. 2002. The SPOT-synthesis technique. Synthetic peptide arrays on membrane supports – principles and applications. *J. Immunol. Methods* 267 (1):13-26.

Gao, Rong, and Ann M. Stock. 2009. Biological Insights from Structures of Two-Component Proteins. *Annual Rev. Microbiol.* 63 (1):133-154.

Geertsma, E.R., N.A.B.N. Mahmood, G.K. Schuurman-Wolters, and B. Poolmann. 2008. Membrane reconstitution of ABC transporters and assays of translocator function. *. Nature Prot.* 3 (2):256-266.

Grote, Mathias, Enrica Bordignon, Yevhen Polyhach, Gunnar Jeschke, Heinz-Jürgen Steinhoff, and Erwin Schneider. 2008. A Comparative Electron Paramagnetic

Resonance Study of the Nucleotide-Binding Domains Catalytic Cycle in the Assembled Maltose ATP-Binding Cassette Importer. *Biophys. J.* 95 (6):2924-2938.

Grote, Mathias, Yevhen Polyhach, Gunnar Jeschke, Heinz-Jürgen Steinhoff, Erwin Schneider, and Enrica Bordignon. 2009. Transmembrane Signaling in the Maltose ABC Transporter MalFGK2-E. *J. Biol. Chem.* 284 (26):17521-17526.

Hänelt, Inga, Dorith Wunnicke, Meike Müller-Trimbusch, Marc Vor der Brüggen, Inga Kraus, Evert P. Bakker, and Heinz-Jürgen Steinhoff. 2010. Membrane Region M2C2 in Subunit KtrB of the K+ Uptake System KtrAB from Vibrio alginolyticus Forms a Flexible Gate Controlling K+ Flux. *J. Biol. Chem.* 285 (36):28210-28219.

Heermann, Ralf, and Kirsten Jung. 2010. Stimulus perception and signaling in histidine kinases. In *Bacterial Signaling*: Wiley-VCH Verlag GmbH & Co. KGaA.

Heermann, Ralf, Arnim Weber, Bettina Mayer, Melanie Ott, Elisabeth Hauser, Günther Gabriel, Torsten Pirch, and Kirsten Jung. 2009. The Universal Stress Protein UspC Scaffolds the KdpD/KdpE Signaling Cascade of Escherichia coli under Salt Stress. *J. Mol. Biol.* 386 (1):134-148.

Hodnik, V, and G. Anderluh. 2010. Capture of intact liposomes on biacore sensor chips for protein-membrane interaction studies. *Methods Mol Biol.* 627 (201-211).

Hoffmann, C., G. Gaietta, M. Bünemann, S.R. Adams, S. Oberdorff-Maass, B. Behr, J.-P. Viladarga, R. Y. Tsien, M. H. Ellisman, and M. J. Lohse. 2005. A FlAsH-based FRET approach to determine G protein–coupled receptor activation in living cells. *Nat Methods* 2 (3):171-176.

Holterhues, Julia, Enrica Bordignon, Daniel Klose, Christian Rickert, Johann P Klare, Swetlana Martell, Lin Li, Martin Engelhard, and Heinz-Jürgen Steinhoff. 2011. The Signal Transfer from the Receptor NpSRII to the Transducer NpHtrII Is Not Hampered by the D75N Mutation. *Biophys. J.* 100 (9):2275-2282.

Hoopmann, MR, CR Weisbrod, and JE Bruce. 2010. Improved strategies for rapid identification of chemically cross-linked peptides using protein interaction reporter technology. *J. Proteome Res.* 9 (12):6323-6333.

Hubsman, Monika, Guennady Yudkovsky, and Ami Aronheim. 2001. A novel approach for the identification of protein.protein interaction with integral membrane proteins. *Nucleic Acids Res.* 29 (4):e18.

Hunke, S., M. Mourez, M. Jéhanno, E. Dassa, and E. Schneider. 2000. ATP modulates subunit-subunit interactions in an ATP-binding-cassette transporter (MalFGK2) determined by site-directed chemical cross-linking. . *J. Biol. Chem.* 275 (15526-15534).

Hunke, Sabine, and Erwin Schneider. 1999. A Cys-less variant of the bacterial ATP binding cassette protein MalK is functional in maltose transport and regulation. *FEBS Lett.* 448 (1):131-134.

Jensen, A.A., J.L. Hansen, S.P. Sheikh, and H. Bräuner-Osborne. 2002. Probing intermolecular protein-protein interactions in the calcium-sensing receptor homodimer using bioluminescence resonance energy transfer (BRET). *Eur J Biochem.* 269 (20):5076-5087.

Johnsson, N, and A Varshavsky. 1994. Split ubiquitin as a sensor of protein interactions in vivo. *Proc. Natl. Acad. Sci. USA* 91 (22):10340-10344.

Johnsson, N, and A. Varshavsky. 1994. Ubiquitin-assisted dissection of protein transport across membranes. *EMBO J.* 13 (11):2686-2698.

Jones, DT. 1998. Do transmembrane protein superfolds exist? . *FEBS Lett* 423:281-285.

Jordan, Patrick, Petra Fromme, Horst Tobias Witt, Olaf Klukas, Wolfram Saenger, and Norbert Krausz. 2001. Three-dimensional structure of cyanobacterial photosystem I at 2.5[thinsp][angst] resolution. *Nature* 411 (6840):909-917.

Jung, Kirsten, Britta Tjaden, and Karlheinz Altendorf. 1997. Purification, Reconstitution, and Characterization of KdpD, the Turgor Sensor of Escherichia coli. *J. Biol. Chem.* 272 (16):10847-10852.

Jura, Natalia, Xuewu Zhang, NicholasÂ F Endres, MarkusÂ A Seeliger, Thomas Schindler, and John Kuriyan. 2011. Catalytic control in the EGF receptor and its connection to general kinase regulatory mechanisms. *Mol. Cell* 42 (1):9-22.

Kaback, H., Irina Smirnova, Vladimir Kasho, Yiling Nie, and Yonggang Zhou. 2011. The Alternating Access Transport Mechanism in LacY. *J. Mem. Biol.* 239 (1):85-93.

Karimova, Gouzel, Nathalie Dautin, and Daniel Ladant. 2005. Interaction network among *Escherichia coli* membrane proteins involved in cell division as revealed by bacterial two-hybrid analysis. *J. Bacteriol.* 187 (7):2233-2243.

Karimova, Gouzel, Josette Pidoux, Agnes Ullmann, and Daniel Ladant. 1998. A bacterial two-hybrid system based on a reconstituted signal transduction pathway. *Proc. Natl. Acad. Sci. U.S.A.* 95 (10):5752-5756.

Klare, J.P. 2012. Site-directed Spin Labeling and Electron Paramagnetic Resonance (EPR) Spectroscopy: A Versatile Tool to Study Protein-Protein Interactions. *InTech*.

Klare, J.P., and H.-J. Steinhoff. 2010. Site-directed spin labeling and pulse dipolar electron paramagnetic resonance. *Encyclopedia of Analytical Chemistry*

Klare, Johann P., Enrica Bordignon, Martin Engelhard, and Heinz-Jürgen Steinhoff. 2011. Transmembrane signal transduction in archaeal phototaxis: The sensory rhodopsin II-transducer complex studied by electron paramagnetic resonance spectroscopy. *Euro. J. Cell Biol.* 90 (9):731-739.

Knierim, Bernhard, Klaus Peter Hofmann, Wolfgang Gärtner, Wayne L. Hubbell, and Oliver P. Ernst. 2008. Rhodopsin and 9-Demethyl-retinal Analog. *J. Biol. Chem.* 283 (8):4967-4974.

Knol, Jan, Klaas Sjollema, and Bert Poolman. 1998. Detergent-Mediated Reconstitution of Membrane Proteins†. *Biochem.* 37 (46):16410-16415.

Krause, Frank. 2006. Detection and analysis of protein–protein interactions in organellar and prokaryotic proteomes by native gel electrophoresis: (Membrane) protein complexes and supercomplexes. *Electrophoresis* 27 (13):2759-2781.

Krogh, Anders, Björn Larsson, Gunnar von Heijne, and Erik L. L. Sonnhammer. 2001. Predicting transmembrane protein topology with a hidden markov model: application to complete genomes. *J. Mol. Biol.* 305 (3):567-580.

Ladant, Daniel, and Gouzel Karimova. 2000. Genetic systems for analyzing protein-protein interactions in bacteria. *Res. Microbiol.* 151 (9):711-720.

Lalonde, Sylvie, David W. Ehrhardt, Dominique Loqué, Jin Chen, Seung Y. Rhee, and Wolf B. Frommer. 2008. Molecular and cellular approaches for the detection of protein–protein interactions: latest techniques and current limitations. *Plant J.* 53 (4):610-635.

Lam, Alice D., Sahar Ismail, Ray Wu, Ofer Yizhar, Daniel R. Passmore, Stephen A. Ernst, and Edward L Stuenkel. 2010. Mapping Dynamic Protein Interactions to Insulin Secretory Granule Behavior with TIRF-FRET. *Biophys. J.* 99 (4):1311-1320.

Locher, Kaspar P., Allen T. Lee, and Douglas C. Rees. 2002. The *E. coli* BtuCD structure: A framework for ABC transporter architecture and mechanism. *Science* 296 (5570):1091-1098.

Magliery, T.J., C.G.M. Wilson, W.L. Pan, and D. Mishler. 2005. Detecting protein–protein interactions with a green fluorescent protein fragment reassembly trap: scope and mechanism. *J. Am. Chem. Soc.* 127:146-157.

Masi, A, R Cicchi, A Carloni, FS Pavone, and A. Arcangeli. 2010. Optical methods in the study of protein-protein interactions. . *Adv Exp Med Biol* 674:33-42.

Mattheyses, Alexa L., Sanford M. Simon, and Joshua Z. Rappoport. 2010. Imaging with total internal reflection fluorescence microscopy for the cell biologist. *J. Cell Sci.* 123 (21):3621-3628.

Maynard, Jennifer A., Nathan C. Lindquist, Jamie N. Sutherland, Antoine Lesuffleur, Arthur E. Warrington, Moses Rodriguez, and Sang-Hyun Oh. 2009. Surface plasmon resonance for high-throughput ligand screening of membrane-bound proteins. *Biotech. J.* 4 (11):1542-1558.

Michnick, Stephen W., Po Hien Ear, Christian Landry, Mohan K. Malleshaiah, and Vincent Messier. 2011. Protein-fragment complementation assays for large-scale analysis, functional dissection and dynamic studies of protein–protein interactions in living cells *Meth. Mol. Biol.* 756 (6):395-425.

Miernyk, Jan A., and Jay J. Thelen. 2008. Biochemical approaches for discovering protein–protein interactions. *Plant J.* 53 (4):597-609.

Müller, Volker Steffen, Peter Ralf Jungblut, Thomas F. Meyer, and S. Hunke. 2011. Membrane-SPINE: An improved method to identify protein-protein interaction partners of membrane proteins *in vivo. Proteomics* 11 (10):2124-2128.

North, A.J. 2006. Seeing is believing? A beginners' guide to practical pitfalls in image acquisition. *J. Cell Biol.* 172:9-18.

Pannier, M., S. Veit, A. Godt, G. Jeschke, and H.W. Spiess. 2000. Dead-time free measurement of dipole-dipole interactions between electron spins. . *J. Magn. Res.* 142 (331-340).

Paternostre, MT, M Roux, and JL. Rigaud. 1988. Mechanisms of membrane protein insertion into liposomes during reconstitution procedures involving the use of detergents. 1. Solubilization of large unilamellar liposomes (prepared by reverse-phase evaporation) by triton X-100, octyl glucoside, and sodium cholate. *Biochem.* 27 (8):2668-2677.

Pelletier, Joelle N., F.-X. Campbell-Valois, and Stephen W. Michnick. 1998. Oligomerization domain-directed reassembly of active dihydrofolate reductase from rationally designed fragments. *Proc. Natl. Acad. Sci. U.S.A.* 95 (21):12141-12146.

Peltan, I.D., A.V. Thomas, I. Mikhaienko, D.K. Strickland, B.T. Hymann, and von Arnim C.A.F. 2006. Fluorescence lifetime imaging microscopy (FLIM) detects stimulus-dependent phosphorylation of the low density lipoprotein receptor-related protein (LRP) in primary neurons. . Biochem. Biophys. Res. Commun. 349:34-30.

Petitjean, A, F Hilger, and K. Tatchell. 1990. Comparison of thermosensitive alleles of the CDC25 gene involved in the cAMP metabolism of Saccharomyces cerevisiae. Gene 124 (4):797-806.

Pfleger, K.D., and K.A. Eidne. 2006. Illuminating insights into protein-protein interactions using bioluminescence resonance energy transfer (BRET). Nat Methods 3 (3):165-174.

Puig, O. , F. Caspary, G. Rigaut, B. Rutz, E. Bouveret, E. Bragado-Nilsson, M. Wilm, and B. Seraphin. 2001. The tandem affinity purification (TAP) method: a general procedure of protein complex purification. Methods 24:218-229.

Reimer, U., U. Reineke, and J Schneider-Mergener. 2002. Peptide arrays: from macro to micro. Curr. Opin. Biotechnol. 13 (4):315-320.

Remy, Ingrid, F. X. Campbell-Valois, and Stephen W. Michnick. 2007. Detection of protein-protein interactions using a simple survival protein-fragment complementation assay based on the enzyme dihydrofolate reductase. Nat. Protocols 2 (9):2120-2125.

Remy, Ingrid, and Stephen W. Michnick. 2007. Application of protein-fragment complementation assays in cell biology. Bio Tech 42 (2):137-145.

Rigaud, J.-L. 2002. Membrane proteins: functional and structural studies using reconstituted proteoliposomes and 2-D crystals. Brazil. J. Med. Biol. Res. 35:753-766.

Rigaud, J.L., and D. Levy. 2003. Reconstitution of membrane proteins into liposomes. . Methods Enzymol. 372 (65-86).

Rigaud, J.L., B. Pitard, and D. Levy. 1995. Reconstitution of membrane proteins into liposomes: application to energy-transducing membrane proteins. Biochim. Biophys. Acta 1231:223-246.

Robichon, Carine, Gouzel Karimova, Jon Beckwith, and Daniel Ladant. 2011. Role of leucine zipper motifs in association of the Escherichia coli cell division proteins FtsL and FtsB. J. Bacteriol. 193 (18):4988-4992.

Saeed, Ramazan, and Charlotte Deane. 2008. An assessment of the uses of homologous interactions. Bioinformatics 24 (5):689-695.

Schäferling, M, and S. Nagl. 2011 Förster resonance energy transfer methods for quantification of protein-protein interactions on microarrays. Methods Mol Biol. 723:303-320.

Schägger, Hermann, and G. von Jagow. 1991. Blue native electrophoresis for isolation of membrane protein complexes in enzymatically active form. Anal. Biochem. 199 (2):223-231.

Shiota, Takuya, Hide Mabuchi, Sachiko Tanaka-Yamano, Koji Yamano, and Toshiya Endo. 2011. In vivo protein-interaction mapping of a mitochondrial translocator protein Tom22 at work. Proc. Natl. Acad. Sci. U.S.A 108 (37):15179-15183.

Silvius, JR. 1992. Solubilization and functional reconstitution of biomembrane components. *Annu. Rev. Biophys. Biomol. Struct.* 21 (323-348).

Skerker, Jeffrey M., Barrett S. Perchuk, Albert Siryaporn, Emma A. Lubin, Orr Ashenberg, Mark Goulian, and Michael T. Laub. 2008. Rewiring the specificity of two-component signal transduction systems. *Cell* 133 (6):1043-1054.

Sun, Y., R.N. Day, and A. Periasamy. 2011. Investigating protein-protein interactions in living cells using fluorescence lifetime imaging microscopy. *Nat Protoc* 6 (9):1324-1340.

Tang, Xiaoting, and James E. Bruce. 2010. A new cross-linking strategy: protein interaction reporter (PIR) technology for protein-protein interaction studies. *Mol. BioSyst.* 6 (6):939-947.

Vaidyanathan, Ravi, Steven M. Taffet, Karen L. Vikstrom, and Justus M. B. Anumonwo. 2010. Regulation of cardiac inward rectifier potassium current (IK1) by synapse-associated protein-97. *J. Biol. Chem.* 285 (36):28000-28009.

Volkmer, Rudolf. 2009. Synthesis and application of peptide arrays: quo vadis SPOT technology. *Chem Bio Chem* 15 (9):1431-1442.

Waters, J.C. 2009. Accuracy and precision in quantitative fluorescence microscopy. *J. Cell Biol.* 185:1135-1148.

Wenschuh, Holger , Rudolf Volkmer-Engert, Margit Schmidt, Marco Schulz, Jens Schneider-Mergener, and Ulrich Reineke. 2000. Coherent membrane supports for parallel microsynthesis and screening of bioactive peptides. *Peptide Science* 55 (3):188-206.

Wittig, Ilka, M Karas, and Hermann Schägger. 2007. High resolution clear native electrophoresis for in-gel functional assays and fluorescence studies of membrane protein complexes. *Mol. Cell. Proteomics* 6 (4):1215-1225.

Wittig, Ilka, and Hermann Schägger. 2008. Features and applications of blue-native and clear-native electrophoresis. *Proteomics* 8 (19):3974-3990.

Repeated Author. 2009. Native electrophoretic techniques to identify protein–protein interactions. *Proteomics* 9 (23):5214-5223.

Wouters, F.S. 2006. The physics and biology of fluorescence microscopy in the life sciences. *Contemp. Phys.* 47:239-255.

Wu, Y, Q Li, and X.Z. Chen. 2007. Detecting protein-protein interactions by Far western blotting. *Nat Protoc* 2 (12):3278-3284.

Xia, Z., and J. Rao. 2009. Biosensing and imaging based on bioluminescence resonance energy transfer. *Curr Opin Biotechnol* 20 (1):37-44.

Xu, Xiaoli, Yuan Song, Yuhua Li, Jianfeng Chang, Hua zhang, and Lizhe An. 2010. The tandem affinity purification method: An efficient system for protein complex purification and protein interaction identification. *Protein Expr. Purif.* 72 (2):149-156.

Yang, Li, Xiaoting Tang, Chad R. Weisbrod, Gerhard R. Munske, Jimmy K. Eng, Priska D. von Haller, Nathan K. Kaiser, and James E. Bruce. 2010. A Photocleavable and Mass Spectrometry Identifiable Cross-Linker for Protein Interaction Studies. *Anal.Chem.* 82 (9):3556-3566.

Yildirim, M. A., K. I. Goh, M. E. Cusick, A. L. Barabasi, and M. Vidal. 2007. Drug-target network. *Nat. Biotechnol.* 25 (1119-1126).

Zhou, Xiaohui, Rebecca Keller, Rudolf Volkmer, Norbert Krauß, Patrick Scheerer, and Sabine Hunke. 2011. Structural insight into two-component system inhibition and pilus sensing by CpxP. *J. Biol. Chem.* 286:9805-9814.

4

Relating Protein Structure and Function Through a Bijection and Its Implications on Protein Structure Prediction

Marco Ambriz-Rivas[1], Nina Pastor[2] and Gabriel del Rio[1]
[1]*Universidad Nacional Autónoma de México, Instituto de Fisiología Celular*
[2]*Universidad Autónoma del Estado de Morelos, Facultad de Ciencias*
México

1. Introduction

Proteins are studied by measuring different properties, typically the chemical structure and biochemical activity. Given that these measurements are done on the same protein molecule, they must be related. Despite the fact that this relationship exists, the mathematical nature of this relationship has remained elusive to our understanding, and is not commonly considered in the so called "structure-function relationship problem of proteins" (Punta & Ofran, 2008). While this is a fundamental problem in biochemistry and biology, that is, to establish a procedure that allows scientists to reliably relate protein structure and protein activity, the likelihood to succeed in this enterprise depends on our ability to understand the mere nature of this relationship. The possibility to effectively relate structure and activity has motivated years of research in different areas in biology, including biophysics, molecular biology, biochemistry, bioinformatics, and computational biology, among others. Although great advances have been achieved from these different areas of expertise, the question remains unsolved. That is, there is no general procedure that may have proven to effectively relate protein structure and activity. However, recent results in the prediction of protein three-dimensional structure (from now on referred simply as 3D structure) are addressing this problem with a fresh look, revealing a new aspect of this relationship that may explain why this particular problem has remained elusive. The present work reviews the general concepts being used to predict protein 3D structure with emphasis on the contribution of these methods to unravel the structure-activity relationship of proteins.

We divide this review in three sections. In the first section, we will present a mathematical view on the evolution of the concept about the 3D structure-activity relationship in proteins. The second section presents the general concepts behind template-based modelling and *ab initio* methods for the prediction of protein 3D structure. There, we will describe how these approaches have contributed to our current understanding of the 3D structure-activity relationship of proteins. Finally, we will review new methods for protein 3D structure prediction and how these may contribute to unravel the 3D structure-activity relationship of proteins.

2. Evolution of the 3D structure-activity paradigm from a mathematical perspective

Back in 1936 Mirsky and Pauling (Mirsky & Pauling, 1936) proposed that protein activity, or its function within a biological context, should be determined by its 3D structure. Considering that the characterization of protein activity has frequently been cumbersome, the possibility to determine it by simply looking at the 3D structure of proteins could be considered an impulse to establish this relationship. Yet, determination of protein 3D structure has not been an easy treat either. Perhaps the main motivation to establish this relationship consists in the possibility to design new devices capable of reproducing the highly efficient capabilities of proteins (Drexler, 1994; Robson, 1999; Balzani *et al.*, 2000) or to simply engineer proteins in order to adapt these for industrial use (Zaks, 2001; Huisman & Gray, 2002; Straathof *et al.*, 2002; Luetz *et al.*, 2008). Ultimately, establishing the 3D structure-activity relationship of proteins may serve to test our level of understanding of these molecules.

Hitherto, the approximation most frequently used to solve this relationship is to consider knowledge-based classification schemes. Such schemes are based on the existence of a given set of proteins with known activity; from that knowledge, it has been possible to identify new proteins sharing similar activity, from protein sequence comparisons. Although quite useful to classify the ever-increasing number of new protein sequences generated nowadays, this type of approaches has a limited ability to assist researchers in the design of protein activity (see next section). Alternatively, the activity of a protein is commonly analyzed from the knowledge of its 3D structure using biophysical methods (Neet & Lee, 2002; Chollet & Turcatti, 1999). In either case, previous knowledge of both protein 3D structure and activity is required to establish this relationship, indicating our current limitation in understanding this problem from basic principles. Even when new enzymatic activities have been designed "from scratch" (Siegel *et al.*, 2010) the active site residues are nestled within previously known protein folds. It has been possible to design completely novel folds, such as Top7, from scratch (Kuhlman *et al.*, 2003), but this refers to the sequence-3D structure relationship, which is not the main focus of this review.

We propose that one of the reasons for this limited understanding of the 3D structure-activity relationship of proteins is the absence of knowledge as to what type of mathematical relationship this one is. As we will show, determining the nature of this relationship may lead researchers to analyze this relationship with a new perspective and may accelerate the full understanding of it.

To explain this, let us first formally describe the 3D structure-activity relationship of proteins as a postulate:

P1: Protein activity depends on its 3D structure.

That is, protein activity may be represented as a mathematical relation of the protein 3D structure. Since both activity and 3D structure can always be measured on a given protein, that is they come in pairs, we postulate that this relation may be represented by a mathematical function. To further describe this postulate, let us define:

D1: Protein activity is defined as the capacity of proteins to interact with other molecules resulting in a change (on the interacting molecule or the environment) that is measurable (e.g., the chemical transformation of glucose to glucose 6-phosphate).

D2: Protein 3D structure is defined by two sets: the set of amino acid residues included in the protein and the set of physical interactions between these residues in the 3D space.

D3. A mathematical function is a particular class of relation between sets and it describes the dependence between the elements of these sets: an independent variable (an element in one of the sets) and the dependent variable (another element in the other set). In other words, for a given value of the independent variable there is one value of the dependent variable.

Postulate **P1** then refers to a mathematical function between two features of proteins: the activity and the 3D structure. The activity is usually expressed as a quantity (kinetic constants such as the Michaellis-Menten constant Km) and the structure may be represented by a quantity also, for instance the fold classification; yet, such quantities have not been easily related, so a new set of measurements is needed to evaluate **P1** (see below for a further discussion on this aspect). To do so, the question we want to address first is: what type of mathematical function is this? Basically, there are three types of mathematical functions:

D4: Injections. In mathematics, this refers to one-to-one relations: given two sets S (3D Structure) and A (protein Activity), there is at least one element in S related with one element in A (see Figure 1A and 1B). Therefore, there can be elements of the set A that do not have a matching partner in set S (Figure 1B).

D5: Surjections. This is defined as a mathematical function where given two sets S and A, there is an association of at least one element in S with an element in A (see Figure 1A and 1C). Therefore, there can be elements of the set A that have one or more relations with elements in set S.

D6: Bijections. These are defined as mathematical functions where for every element in set S there is exactly one element in set A associated to it. They occur when both an injection and a surjection relation exist (see Figure 1A).

In all these cases (injections, surjections and bijections), the mathematical function f might be reversible: given f: S → A, then it is possible to find a function g such that g: A → S. However, only in the case of bijections the reversibility of the association is a necessary condition of the function.

Expressing these concepts in terms of the 3D structure-activity relationship of proteins, we may say that this relationship presents the properties of injections. For a long time biochemists have characterized the activities of proteins; however, for some time many activities were known but no protein 3D structures were associated to them; more recently, with the advent of DNA sequencing, many protein sequences and 3D structures are known for which no activity has being assigned yet (Norin & Sundström, 2002). However, given postulate P1, we must expect that for each protein there have to be both an activity and a 3D structure associated to it; consequently, the currently unknown 3D structures or activities of proteins will be measured eventually.

Alternatively, most of the current approaches to study the 3D structure-activity relationship of proteins treat this as a surjection: the evolution theory postulates that protein activity or 3D structure has been conserved in different species (orthologous proteins); thus this is a case of a one-to-many (one function-many structures) relation. Additionally, in protein evolution the term "convergence" refers to the cases where different 3D structures of proteins have evolved to share a similar activity; conversely, an alternative example are single-domain moonlighting proteins, where one 3D structure is associated with multiple activities, albeit, using different molecular surfaces (Jeffery, 1999, 2003, 2009; Copley, 2003). In any case though, the one-to-many association prevails as much as we group together 3D structures or activities that are not identical. That is, to the best of our knowledge, there are no two proteins with identical activities reported so far with perfectly different 3D structures, nor are there two proteins with identical 3D structures with perfectly different activities. Take for instance the triose-phosphate isomerase proteins; these are proteins with a high degree of sequence-3D structure similarity, sharing similar but not identical activities (see Table 1). In the case of moonlighting proteins, there is no evidence that the two different activities may be performed in the same protein having exactly the same 3D structure, yet the structure may be slightly altered to accomplish different activities (Bateman *et al.*, 2003; Krojer *et al.*, 2002).

The need to move from considering similar to identical activity or 3D structure in the structure-activity relationship of proteins is important to improve our understanding of this relationship. On the one hand, it is convenient to assume similarity in 3D structure or activity of proteins in the discovery phase of biology (i.e., accelerated discovery of new proteins) because this assumption allows for the classification of new proteins into known families of proteins with known activity. Alternatively, provided the existence of an activity assay, it is possible to identify new proteins with such activity and presumably related in their 3D structure. However, after this initial phase of discovery, full understanding of the activity or 3D structure of a protein requires more detailed analysis both experimentally and theoretically. For the theoretical part, here we claim that in order to gain a better understanding of the 3D structure-activity relationship of proteins it is necessary to be precise in the terms used to relate these properties.

From this analysis we noted that since the 3D structure-activity relationship of proteins presents features of both injections and surjections, thus it may be best represented by a bijection. Furthermore, assuming that the injective feature is only a temporal one, and the surjective feature exists if and only if the definition of activity or 3D structure is not precise, we may conclude that the best way to analyze the 3D structure-activity relationship of proteins is as a bijection, where we postulate that for any given protein there is always one activity related to a given 3D structure. This approach necessarily implies that one has to come up with a rigorous and precise definition for both 3D structure and activity. Herein lies the challenge.

This conclusion leads us to the following scenario: let us assume that there is a set S with every possible 3D structure of proteins, and a set A with every possible measurable activity of proteins; then, for a given protein 3D structure in set S there is exactly one protein activity in set A; conversely, for a given protein activity in set A, there is exactly one protein 3D structure in set S. In this scenario, there are no identical activities in set A, neither there are identical structures in set S. To formally express this:

$$A = f(S) \tag{1}$$

Now, in order to express this relation in numerical terms, let us define the 3D structure as a matrix (e.g., adjacency matrix) and activity as a vector (e.g., list of critical residues for protein activity). Choosing this set of critical residues is a convenient pick since it has been reported that proteins sharing high 3D structural similarity do not share the same set of critical residues (Cota E *et al.*, 2000; Rivera MH *et al.*, 2003), yet some critical residues are indeed shared between homologue proteins (Zhang Z & Palzkill T, 2003). Thus, representing 3D structure as a matrix (M) and activity as a vector of critical residues (C) provides us with a way to express this relation formally and look for mathematical tools to define the mathematical function inherent to these quantities. Thus:

$$C = f(M) \qquad (2)$$

In other words, given a set of contacts between the residues of a protein (3D structure), our problem is to find a mathematical transformation of this matrix into a vector containing the critical residues for the protein function. If the mathematical function relating M and C is a bijection, then it must be possible to transform the vector C back into the matrix M. In order to find the mathematical function involved in this transformation, having access to multiple 3D structures and multiple sets of critical residues for several proteins is required.

Our analysis has several implications for the analysis of the 3D structure-activity relationship. In the present review, we will discuss only those relevant for the prediction of protein 3D structure. That is encouraged by the emergence of new approaches for the prediction of protein 3D structure that are based on the notion that the 3D structure-activity relationship is a bijection. However, these approaches have been developed in the absence of the current mathematical context, as we will describe below; embracing this bijection may provide the basis to improve the current methods of protein 3D structure prediction.

3. Current methods for protein 3D structure prediction

In this section we will summarize the ideas behind them and the kind of relationship that they assume between 3D structure and activity. This review does not attempt to cover in detail these methodologies, but to present the basic aspects of them in the context of postulate P1. For detailed descriptions of these methodologies, there are other reviews published elsewhere (Jones & Thornton, 1993; Martí-Renom *et al.*, 2000; Osguthorpe, 2000; Hardin *et al.*, 2002; Koretke *et al.*, 2002; Zhang, 2002; Godzik, 2003).

3.1 General considerations

Despite of the diversity of approaches to perform structural predictions, they all share a common design. The two key components of any method are the model generator and a quality evaluator (Figure 2).

1. Model generators refer to algorithms that create native-like protein 3D structures. There are two ways to generate such structures: knowledge-based strategies that depend on the available structures in databases and *ab initio* strategies (also known as physics-based), which consider physics principles to generate structures. Typically, model generators produce many alternative 3D structures that are potential solutions to the native structure of the protein.

2. Quality evaluators. These algorithms aim to evaluate the quality of the models produced by the model generators, in order to select the best models; i.e., those resembling the known native-like structure of proteins. Like the model generators, quality evaluators can be knowledge-based or *ab intio*.

It is important to keep in mind that these methodologies have limitations, especially if they are used to gain insights into the relation between the 3D structure and activity of poorly characterized proteins. Knowledge-based model generators and evaluators assume surjective relations between structure and activity, since the common idea of modellers of protein 3D structures is to assist in the grouping of protein structures based on similar attributes (Gerstein & Hegyi, 1998; Domingues *et al.*, 2000; Skolnick *et al.*, 2000). Therefore, in these cases knowledge of the protein 3D structure may provide inaccurate information about the activity (Martin *et al.*, 1998). On the other hand, *ab initio* methods do not take into account the 3D structure-activity relation to perform predictions. With this kind of predictions, it is unlikely to get precise information about the activity of the protein from its 3D structure (Baker & Sali, 2001, and the results from CASP9).

Often the 3D structure is used to interpret the activity and rarely the other way around (Gherardini & Helmer-Citterich, 2008), thus it is not surprising that the current methods of protein 3D structure prediction do not address the prediction of 3D structure from the activity of the protein. In spite of this limitation, current methodologies for protein 3D structure prediction have been important in the development of the ideas about protein 3D structure determinants and their relationship with activity. Consequently, in the next two sections we will describe briefly the current methods for protein structure predictions, their features and limitations to elucidate protein activity.

3.2 Template-based modeling

This kind of predictions uses a protein of known 3D structure as a template to build the model of a protein whose 3D structure is unknown (target). The most critical part of this methodology is to identify adequate template(s) for the target. Accordingly, template-based modelling is classified in two main areas: homology modelling and fold recognition.

The idea behind homology modelling is that similar sequences have similar 3D structures (Doolittle, 1981, 1986; Chothia & Lesk, 1986). In this regard, the quality of a 3D model for a target protein depends strongly on the percentage of sequence identity between the target and template; the greater the identity, the more accurate the model will be. Likewise, below 30% of identity between the target and template proteins (sometimes referred as the "twilight zone"; Doolittle, 1986), several false templates may be identified for the target protein (Sander & Schneider, 1991; Rost, 1999). In that case, templates should be searched with fold recognition algorithms (Rost, 1999; see below). Templates can be found by searching databases of proteins with known 3D structure (e.g. the Protein Data Bank) with sequence alignment tools like BLAST (Altschul *et al.*, 1990, 1997) or FASTA (Pearson & Lipman, 1988). Then, models of the target protein are built from the templates, taking into account changes that must be introduced like insertions and deletions in the template (indels), side chain conformations of non-conserved residues, possible rearrangements in the backbone, among others (Jones & Thirup, 1986; Bruccoleri & Karplus, 1987; Vásquez, 1996).

Afterwards, the quality of the resulting models is evaluated (Laskowski *et al.*, 1993; Hooft *et al.*, 1996; Wallner & Elofsson, 2003; Ginalski *et al.*, 2003).

On the other hand, fold recognition methodologies identify proteins sharing similar 3D structures even if they do not have any obvious sequence similarity (Jones & Thornton, 1993; Godzik, 2003). Fold recognition can be performed in two ways. The first involves the enhancement of homology detection (Fischer & Eisenberg, 1996; Jaroszewski *et al.*, 1998; Rychlewski *et al.*, 2000), by using sequence profiles compiled from protein sequences that are compatible with the target. Two examples of this approach are PSI-BLAST (Altschul, 1997) and hidden Markov models (Durbin *et al.*, 1998). Accuracy of prediction is increased further if structural information (e.g. secondary structure) is incorporated in the profiles (Di Francesco *et al.*, 1997a, 1997b). The second approach is termed "threading" (Jones *et al.* 1992; Godzik & Skolnick, 1992). Here, the target sequence is forced to adopt the 3D structure of a potential target. Then the quality of the model is evaluated with a structure-based score. If the model has a high score, there is confidence that the target adopt a similar 3D structure as the template, otherwise the model is discarded. Once the template(s) is (are) found, a 3D-structural model of the target protein is built following the steps described in homology modelling after the initial template identification.

Template-based modelling has been recognized as the most accurate approach for protein structure prediction, especially if the identity between target and template is high (Chothia & Lesk, 1986; Sali *et al.*, 1995; Cozzetto *et al.*, 2009). However, as any model, these need to be tested in their ability to reproduce a biologically relevant feature, such as the activity. Since these methods assume a surjection for the structure-activity relationship, there are limitations imposed by such assumption, which are more notorious in the cases of low sequence similarity between the target and template proteins. One example of the limitation induced by the surjection conjecture in the structure-activity relationship of proteins is the TIM barrel fold, a common 3D-structure present in enzymes with very different activities such as oxidoreductases, hydrolases, lyases and isomerases (Greene *et al.*, 2007). Likewise, the opposite situation is common: proteins with very similar activities and structurally unrelated. For instance, both chymotrypsin and subtilisin are serine-proteases with the same catalytic triad in the active site even thought they have completely different 3D-structures (Wallace *et al.*, 1996).

Furthermore, even when there is a clear similarity between target and template sequences, there can be measurable structural differences. The most common example is loop structure. Precise prediction of loop regions is usually hard to accomplish since they tend to exhibit higher sequence variability and often have insertions and deletions relative to templates (Martí-Renom *et al.*, 2000). Loops though, play an important role conferring specificity to the protein activity. Another less frequent situation is when there are visible differences in active sites of related proteins. This can lead to inaccurate modelling of the structure of target proteins (Moult, 2005). One way to improve the modelling of loops would be to evaluate the predicted activity of the model.

The information summarized above provides a general notion about the relationships that template-based modelling assumes. One-to-many relations between protein structure and activity are quite common with this kind of predictions. Thus, it is frequent to misrelate the activity of a protein from the knowledge of its fold alone (Martin *et al.*, 1998). It is often necessary to use other resources to predict the activity more accurately, as the use of local structural features of proteins in active sites (see Gherardini & Helmer-Citterich, 2008 for more details). Such tools work with the traditional approaches for predictions: knowledge-based like the 3D-templates (Wallace *et al.*, 1996); or physics based, for example the identification of clefts and pockets in protein structures (Laskowski *et al.*, 1996; Binkowski *et al.*, 2003). These methods provide a theoretic framework to understand the 3D structure-activity relation in a one-way path: the prediction of activity from structure.

3.3 *Ab initio* modeling

Template based modelling can provide insights into the 3D structure and activity of poorly characterized proteins. In terms of generating reliable models it has an intrinsic limitation: it requires a protein of known 3D structure in order to produce a model. This may not be a problem in many situations, but there are proteins without any detectable template (more than half of the sequenced proteins in known genomes, see Yura *et al.*, 2006). In such cases the alternative is *ab initio* modelling, also known as template-free modelling (Osguthorpe, 2000; Hardin *et al.*, 2002; Koretke *et al.*, 2002). The premise of these modelling methods is that the protein sequence determines the native structure, which has the global minimum potential energy among all the alternative conformations. In other words, *ab initio* methods assume that sequence alone would be sufficient to model the structure of proteins. For this reason, *ab inito* methods are adjured to predict the structure folds that were previously unknown.

Ab initio methods carry out a large-scale search for protein structures that have a particularly low energy for a given amino acid sequence. The two critical parts of these predictors are the conformational search strategy and the energy evaluation method (known as energy potential). To perform a fast and efficient search of the conformational space, *ab initio* methods use sophisticated algorithms suited to solve combinatorial problems since it is impossible to systematically explore all the conformations of a polypeptide chain. Monte Carlo algorithms (Simons *et al.*, 1999; Ortiz *et al.*, 1999), genetic algorithms (Pedersen & Moult, 1997a, 1997b), zipping and assembly (Ozcan *et al.*, 2007) and molecular dynamics (Duan & Kollman, 1998; Shaw DE *et al.* 2010) are among the most frequently used methods to explore the conformational space of protein structures. Likewise, the energy potential is crucial to evaluate and select models of the target protein. Energy potentials can be of two kinds: molecular mechanics potentials, that are derived from physical-chemical calculations (Brooks *et al.*, 1983; Pearlman *et al.*, 1995) and knowledge-based potentials are constructed from the statistical analysis of the available structures in databases (Sippl, 1990; Koretke *et al.*, 1998; Kuhlman and Baker, 2000).

Ab initio predictions usually consume a great deal of time and computer power. Recent methods make simplifications on the protein 3D structure in order to keep an acceptable speed (Helles, 2007). One of the solutions is to reduce the number of atoms that represent the protein 3D structure in order to simplify the model generation process (Kolinski, 2004; Lee *et al.*, 1999). An alternative to speed up calculations is to consider fragment assembly strategies (Simons *et al.*, 1999; Jones & Thirup, 1986). The idea with this approach is to split the structure into smaller fragments composed by many residues. Fragments are selected from a knowledge-based database on the basis of structural compatibility with the target sequence and secondary structure propensities. The assembly of such substructures is determined by the energy potential and the conformation searching strategy. There are also multi-scale methods, like those of Cecilia Clementi, which change the resolution of the model depending on the questions that want to be asked of the protein (Shehu *et al.*, 2009).

Template-free modelling has experienced much progress since the first blind prediction experiment known as "Critical Assessment of Techniques for Protein Structure Prediction" (CASP) took place in the early 90's (Bourne, 2003; Moult, 2005). However, despite of the considerable efforts the accuracy of *ab initio* predictions is still very low, compared to template-based modelling. That is, models generated with *ab initio* methods may have very large deviations from the experimental structures. In other cases, the 3D structure of the model can be completely wrong (this is actually a common situation). These limitations have hindered the practical use of *ab intio* modeling for the inference of the 3D structure-activity relationship on the target proteins (Baker & Sali, 2001; the results from CASP9).

Finally, *ab initio* predictions do not take into account the relation between 3D structure and activity explicitly, therefore they provide little reliable information about this relationship. On the other hand, they assume that proteins fold autonomously to the 3D structure with the minimum free energy (this is the case for most globular proteins), but there are cases where this assumption may be unjustified, as in the case of protein folding under kinetic control. For example, it has long been recognized that transmembrane proteins do not adopt their final, functional 3D-structure unassisted, but they need a translocation machinery to insert into the membrane and fold (Elofsson & von Heijne, 2007). Hence, the use of these strategies is inadequate for transmembrane proteins. Nonetheless, the ROSETTA method (originally developed for globular proteins) has been adapted to predict transmembrane proteins, with limited success (Yarov-Yarovoy *et al.*, 2006). Additionally, *ab initio* predictions are unsuited for natively unstructured proteins (proteins that do not have a defined, unique structure), because they perform their activities as many alternative, rapidly interchanging conformations that correspond to multiple energy minima (Radivojac *et al.*, 2007).

Despite of these disadvantages, *ab initio* predictions sometimes provide insights about protein activity. For example, in the fourth CASP experiment, the ROSETTA method was able to predict the structure of a couple target proteins that are structurally related to proteins of known 3D structure that were missed by fold recognition methods (Bonneau *et al.*, 2001; Baker & Sali, 2001). Interestingly, the activities of the target proteins were similar, even thought there was no significant sequence identity between the proteins. A second example is the signalling protein Frizzled, whose critical residues for activity (previously characterized) were clustered together in the predicted structure in a surface patch likely to be involved in key protein-protein interactions (Baker & Sali, 2001). From these examples, it can be concluded that *ab initio* methods are more effective to gain information about the activity if they are combined with knowledge-based approaches (carrying on their limitations).

3.4 Concluding remarks about the current methods for protein 3D structure prediction

The available methodologies for the 3D structure prediction of proteins have provided useful insights about the relation between 3D structure and activity, and helped to construct the current paradigm. However, further refinement of these methods may assist to fully relate protein 3D structure with activity. In this review we propose that such refinement may come from the recognition of the bijective nature of the 3D structure-activity relationship. For instance, knowledge-based methods imply a surjective relationship between activity and 3D structure. Consequently, predicting details on the activity of a modelled 3D structure of a protein can be hard, since there are examples of folds associated with many activities and *vice versa*. Furthermore, *ab initio* methods do not consider the structure-activity relationship, therefore the information they provide about the activity is commonly inaccurate. Additionally, template-free methods assume that proteins fold autonomously into a stable, minimum energy conformation, limiting their applicability in proteins that do not have these features because they fold under kinetic control.

In summary, it is necessary to develop methods that take into account the bijective nature of the 3D structure-activity relationship, in order to improve the usefulness and reliability perhaps, of protein 3D structure prediction methods. In the following section we will describe the available methodologies that take into account this bijection.

4. Emerging methods for protein structure prediction based on the bijective nature of protein 3D structure and activity

The previous section outlined the current status in the protein 3D structure prediction field, its strengths and weaknesses with regard to activity inference. It is evident that current methodologies still have limitations to exploit the usefulness of the 3D structure-activity relationship. Fortunately, new methodologies have been developed that take into account the bijective nature between the 3D structure and activity of proteins. This section will discuss the principles behind these methods and their capabilities.

4.1 Relevance of critical residues in the 3D structure-activity relation

These methods are based on the concept of critical residues, which are defined as those residues that upon mutation abolish the activity of a protein. Such definition depends on the experimental procedure used to measure the activity of the protein, but generally speaking, residues are considered critical if they tolerate few if any mutations (Loeb *et al.*, 1989; Rennell *et al.*, 1991; Terwilliger *et al.*, 1994; Huang *et al.*, 1996; Axe *et al.*, 1998). Therefore, an experimentally determined critical residue may be either important to maintain the 3D structure of a protein or critical for the interaction with another molecule, or both. Thus, these residues constitute a key piece of knowledge that can be exploited to relate activity and 3D structure. Not surprisingly, methods have been developed to predict critical residues from protein sequence and/or 3D structure (Elcock, 2001; del Sol Mesa *et al.*, 2003; Glaser *et al.*, 2003; Thibert *et al.*, 2005; Cusack *et al.*, 2007).

Additionally, critical residues may provide a useful way to quantify structural features of proteins and relate them with the activity of a protein. As we mentioned earlier, there are no reports of two proteins with identical 3D structures with perfectly different activities and *vice versa* (please note that correctly representing both 3D structure and activity is one of the biggest challenges, and therefore, a Cartesian representation of the protein may not be the best to distinguish identical 3D structures). Hence, it is expected that proteins with similar, yet strictly different 3D structures, will have different sets of critical residues. If that is the case, the set of critical residues for a given protein should reflect its unique 3D structural and activity properties. Such assumption provides the framework for methodologies that are based in the bijective relation between 3D structure and activity.

In the next two sections, we will describe the available bijective approaches. To simplify, they are classified in two categories: phylogeny and structure-based methods. The usefulness of these methodologies to relate 3D structure and activity will also be discussed.

4.2 Phylogeny-based approaches

The idea behind phylogenetic methods is to exploit the evolutionary information that can be extracted from the analysis of the sequences of related proteins. To do so, it is necessary to identify a group of similar protein sequences and to construct a multiple sequence alignment with them. There are two types of information that can be extracted from the alignments: sequence conservation and sequence correlation.

The first property refers to the frequency of a specific amino acid at a given position in the alignment; residues occurring at high frequencies at particular positions are considered conserved residues. Sequence conservation is related to the direct evolutionary pressure to maintain the physical-chemical characteristics of some positions in order to retain the activity and/or 3D structure of a family of homologous proteins. Therefore, highly conserved residues are regarded as critical to retain the 3D structure and activity of the protein. In the literature, there are many reports of methods to calculate conservation (see Valdar *et al.*, 2002 and Sadowski & Jones, 2009 for comprehensive reviews).

Residue correlation (also known as co-evolution or co-variation) is defined as concerted patterns of variation between two or more different positions in a multiple sequence alignment of homologous proteins (Altschuh *et al.*, 1987). Such co-variating residues are proposed to correspond to compensatory substitutions that maintain the structural stability or functional properties of proteins throughout their evolutionary history. It has been observed that correlated residues tend to be in physical contact (Altschuh *et al.*, 1988); thus, this feature was proposed to be useful in residue contact predictions (Göbel *et al.*, 1994; Pazos *et al.*, 1997; Olmea & Valencia, 1997).

Critical residues predicted with phylogenetic approaches can be exploited to improve structural predictions. For example, the method reported by the group of Valencia (Olmea *et al.*, 1999) uses sequence conservation and correlation as part of a structure quality evaluator for a fold recognition structure predictor. The authors of this work report that the method is capable to distinguish correct models from incorrect models generated by the TOPITS threading algorithm (Rost, 1995). However, the accuracy of the algorithm decreases for large proteins, thus restricting its applicability.

Another exciting application of sequence correlation and co-variation is the design of new proteins (a field that strongly depends on 3D structure prediction tools). An illustrative example of protein engineering is the use of the Statistical Coupling Analysis method (SCA; Lockless & Ranganathan, 1999), which was used to design a novel artificial protein sequence with the same 3D structure and activity as natural WW domain proteins (Socolich *et al.*,

2005; Russ *et al.*, 2005). In order to design the protein, the method took into account the critical residues of the protein as well as their patterns of conservation and correlation (Socolich *et al.*, 2005). Furthermore, the methodology has been used recently to design a light-modulated chimerical enzyme (Lee *et al.*, 2008).

Ultimately, conserved residues will only capture the common critical residues for a set of homologous proteins, and will most likely miss the critical residues specific for the activity and 3D structure of each protein in that set. In that sense, conserved residues may be useful to score common structural features of proteins but may not be useful to evaluate the different 3D structure and biological activity of each homologous protein. To do so, a new method has recently being described that is now reviewed.

4.3 Methods based on structural information

A complementary approach to identify critical residues is to consider only 3D structural properties of proteins. One of the most recent approaches to study the 3D structure of proteins is graph theory (Vendruscolo *et al.*, 2002; Greene & Higman, 2003; Thibert *et al.*, 2005; Cusack *et al.*, 2007; Montiel Molina *et al.*, 2008), a theoretical approximation that has been used to characterize other biological systems, such as metabolism, genetic regulation and protein-protein interaction networks (Jeong *et al.*, 2000, 2001; Del Rio *et al.*, 2009). Under this view, protein 3D structure is modelled as a graph (network), which is defined by one set of nodes that represent the amino acid residues in a protein, and a set of edges that can be considered as molecular interactions between any two residues (nodes). The criterion to link two residues by an edge is based on maximum distances among the atoms of residues (Vendruscolo *et al.*, 2002; Greene & Higman, 2003; Thibert *et al.*, 2005; Cusack *et al.*, 2007; Milenković *et al.*, 2009).

Graph theory provides the mathematical basis to study the topological properties of networks derived from the protein structure. One useful concept of this field to characterize networks is network centrality, which measures the relative importance of nodes in the network. Thus, centrality can be used to predict critical residues (Thibert *et al.*, 2005; Cusack *et al.*, 2007) or to study topological features of protein structures (Vendruscolo *et al.*, 2002; Greene & Higman, 2003). Some of the most common centralities used to study networks derived from protein structures are betweenness and closeness, which relate nodes through the shortest paths among all the nodes in the graph (Freeman, 1977).

Centrality is reliable when it comes to predict critical residues (Chea & Livesay, 2007), but how can these be used to predict 3D structural and functional features? We have recently reported a tool named "JAMMING" to facilitate this task. The method predicts critical residues using betweenness or closeness centrality (Cusack *et al.*, 2007). We have shown that JAMMING may be used to identify protein structures involved in ligand binding by screening thousands of conformations generated from protein 3D structures in the unbound form; such functional conformers were found by a scoring system that matches critical residues with central residues (Montiel Molina *et al.*, 2008). Our results show that critical residues for a molecular interaction are preferentially found as central residues of protein structures in complex with a ligand. Therefore, the tool helps to relate the activity of the protein (binding to a molecule) with its structural properties (the conformers).

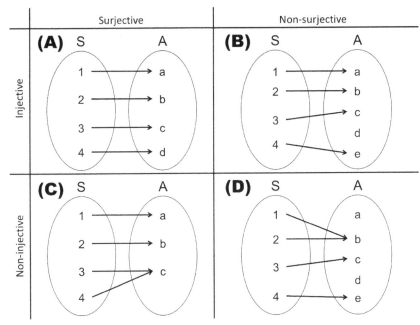

Fig. 1. Examples of injective and surjective functions
A) Injective and surjective (bijection). B) Injective and non-surjective. C) Non-injective and surjective. D) Non-injective and non-surjective.

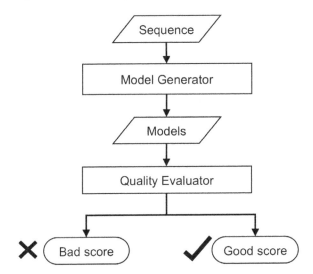

Fig. 2. Flowchart of structural prediction methods. The protein sequence is the input of the model generator algorithm. As a result, the generator produces multiple models that are assessed by the quality evaluator. Finally, the best scoring models are selected, whereas the models with bad scores are discarded.

Species	PDB	Identity[1] [%]	RMSD[1] [A]	K_m [mM]	K_{cat} [1/s]	References[2]
Homo sapiens	1wyi	100	0.0	0.34	16320	Gracy, 1975
Oryctolagus cuniculus	1r2t	98	0.4	0.42	8670	Krietsch, 1975a
Gallus gallus	8tim	89	0.8	0.47	4300	Xiang et al., 2004
Saccharomyces cerevisiae	1ypi	53	1.0	1.27	16700	Krietsch, 1975b
Trypanosoma brucei	1tpf	53	1.1	0.19	6000	Kursula et al., 2002
Leishmania mexicana	1amk	50	1.6	0.30	4170	Kohl et al., 1994
Escherichia coli	1tre	46	1.4	1.03	9000	Alvarez et al., 1998
Vibrio marinus	1aw2	42	1.4	1.90	7000	Alvarez et al., 1998

[1] Sequence identities and RMSDs were calculated with the program DaliLite (Holm & Park, 2000) using 1wyi as the first molecule in all comparisons.
[2] References originally reporting the values for Km and Kcat.

Table 1. Structural and functional features of triose-phosphate isomerases from different species.

5. Conclusions

The structure-activity paradigm has travelled a long way since the first efforts to characterize the 3D structure and biological activity of proteins were performed back in the 1930's. Traditionally, the relationship between 3D structure and activity has been considered as a surjection to assist in the classification of the known proteins. Consequently, knowledge-based classification schemes, although useful to give sense to an ever-increasing list of known protein sequences, may not provide the basis to understand the subtleness of protein activity and structure in nature. In a similar fashion, most of the current methods for protein 3D structure prediction are unable to provide better insights about the activity of a protein of unknown structure (especially if it does not have a close homologue).

In this review, we propose that the relation between structure and activity may be modelled by a bijection. Critical residues provide a way to relate the structure and the activity of proteins, especially in the situation where structure and activity are represented by a bijection. Current methodologies based on the bijective 3D structure-activity relationship unnoticeably provided novel tools to explore the subtle determinants of protein activity, structure and their interaction. We claim that the incorporation of these methods into the traditional tools for protein structure prediction will improve the usefulness of the structural predictions to understand the details on the evolution of protein activity.

6. Acknowledgment

We want to acknowledge the technical assistance of Dra. Maria Teresa Lara Ortiz and the IT core facility of the Instituto de Fisiología Celular; Dr. Alejandro Fernández for his fruitful discussions on this subject and reading of the manuscript. This work was supported in part by one grant from CONACyT (82308) and two grants from the Universidad Nacional Autónoma de México to GDR, including the Macroproyecto: Tecnologías para la Universidad de la Información y la Computación and PAPIIT IN205911, and CONACyT (102182 and 133294) to NP.

7. References

Altschuh D, Lesk AM, Bloomer AC, Klug A. (1987). Correlation of co-ordinated amino acid substitutions with function in viruses related to tobacco mosaic virus. J Mol Biol. 193(4):693-707.

Altschuh D, Vernet T, Berti P, Moras D, Nagai K. (1988). Coordinated amino acid changes in homologous protein families. Protein Eng. 2(3):193-199.

Altschul SF, Gish W, Miller W, Myers EW, Lipman DJ. (1990). Basic local alignment search tool. J Mol Biol. 215(3):403-410.

Altschul SF, Madden TL, Schäffer AA, Zhang J, Zhang Z, Miller W, Lipman DJ. (1997). Gapped BLAST and PSI-BLAST: a new generation of protein database search programs. Nucleic Acids Res. 25(17):3389-402.

Alvarez M, Zeelen JP, Mainfroid V, Rentier-Delrue F, Martial JA, Wyns L, Wierenga RK, Maes D. (1998). Triose-phosphate isomerase (TIM) of the psychrophilic bacterium Vibrio marinus. Kinetic and structural properties. J Biol Chem. 273(4):2199-2206.

Axe DD, Foster NW, Fersht AR. (1998).A search for single substitutions that eliminate enzymatic function in a bacterial ribonuclease. Biochemistry. 37(20):7157-7166.

Baker D, Sali A (2001). Protein structure prediction and structural genomics. Science. 2001. 294(5540):93-6.

Balzani V V, Credi A, Raymo FM, Stoddart JF. (2000). Artificial Molecular Machines. Angew Chem Int Ed Engl39(19):3348-3391.

Bateman OA, Purkiss AG, van Montfort R, Slingsby C, Graham C, Wistow G. (2003). Crystal structure of eta-crystallin: adaptation of a class 1 aldehyde dehydrogenase for a new role in the eye lens. Biochemistry. 42(15):4349-56.

Binkowski TA, Adamian L, Liang J. (2003). Inferring functional relationships of proteins from local sequence and spatial surface patterns. J Mol Biol. 332(2):505-26.

Bonneau R, Tsai J, Ruczinski I, Baker D. (2001). Functional inferences from blind ab initio protein structure predictions. J Struct Biol. 134(2-3):186-90.

Bourne PE. (2003).CASP and CAFASP experiments and their findings. Methods Biochem Anal. 44:501-7.

Brooks BR, Bruccoleri RE, Olafson BD, States DJ, Swaminathan S, Karplus M. (1983). CHARMM: A program for macromolecular energy, minimization, and dynamics calculations. J Comp Chem. 4(2):187–217.

Bruccoleri RE, Karplus M. (1987). Prediction of the folding of short polypeptide segments by uniform conformational sampling. Biopolymers. 26(1):137-168.

Chea E, Livesay DR. (2007). How accurate and statistically robust are catalytic site predictions based on closeness centrality? BMC Bioinformatics. 8:153.

Chollet A, Turcatti G. (1999). Biophysical approaches to G protein-coupled receptors: structure, function and dynamics. J Comput Aided Mol Des. 13(3):209-219

Chothia C, Lesk AM. (1986). The relation between the divergence of sequence and structure in proteins. EMBO J. 5(4):823-6.

Copley SD. (2003). Enzymes with extra talents: moonlighting functions and catalytic promiscuity. Curr Opin Chem Biol. 7(2):265-72.

Cota E, Hamill SJ, Fowler SB, Clarke J. (2000). Two proteins with the same structure respond very differently to mutation: the role of plasticity in protein stability. J Mol Biol. 302(3):713-25.

Cozzetto D, Kryshtafovych A, Fidelis K, Moult J, Rost B, Tramontano A. (2009) Evaluation of template-based models in CASP8 with standard measures. Proteins 77 Suppl 9:18-28.

Cusack MP, Thibert B, Bredesen DE, del Rio G (2007) Efficient identification of critical residues based only on protein structure by network analysis. PLoS ONE 2(5):e421.

Del Rio G, Koschützki D, Coello G (2009) How to identify essential genes from molecular networks? BMC Syst. Biol. 3:102.

del Sol Mesa A, Pazos F, Valencia A (2003) Automatic methods for predicting functionally important residues J Mol Biol. 326(4):1289-1302.

Di Francesco V, Garnier J, Munson PJ. (1997a). Protein topology recognition from secondary structure sequences: application of the hidden Markov models to the alpha class proteins. J Mol Biol. 267(2):446-463.

Di Francesco V, Geetha V, Garnier J, Munson PJ.(1997b). Fold recognition using predicted secondary structure sequences and hidden Markov models of protein folds. Proteins. Supplement 1:123-128.

Domingues FS, Koppensteiner WA, Sippl MJ. (2000). The role of protein structure in genomics. FEBS Lett. 476(1-2):98-102.

Doolittle RF. (1981). Protein Evolution. Science. 214(4525):1123-1124.

Doolittle RF. (1986). Of URFs and ORFs: a primer on how to analyze derived amino acid sequences. University Science Books, Mill Valley, CA, USA.

Drexler KE. (1994). Molecular nanomachines: physical principles and implementation strategies. Annu Rev Biophys Biomol Struct. 23:377-405.

Duan Y, Kollman PA. (1998). Pathways to a protein folding intermediate observed in a 1-microsecond simulation in aqueous solution. Science. 282(5389):740-744.

Durbin R, Eddy S, Krogh A and Mitchison G. (1998). Biological Sequence Analysis: Probabilistic Models of Proteins and Nucleic Acids. Cambridge: Cambridge University Press.

Elcock AH (2001) Prediction of functionally important residues based solely on the computed energetics of protein structure J Mol Biol. 312(4):885-896.

Elofsson A, von Heijne G. (2007). Membrane protein structure: prediction versus reality. Annu Rev Biochem. 76:125-140.

Fischer D, Eisenberg D. (1996). Protein fold recognition using sequence-derived predictions. Protein Sci. 5(5):947-55.

Freeman LC. (1977). A set of measures of centrality based on betweenness. Sociometry 40:35-41.

Gerstein M, Hegyi H. (1998). Comparing genomes in terms of protein structure: surveys of a finite parts list. FEMS Microbiol Rev. 22(4):277-304.

Gherardini PF, Helmer-Citterich M. (2008). Structure-based function prediction: approaches and applications. Brief Funct Genomic Proteomic. 7(4):291-302.

Ginalski K, Elofsson A, Fischer D, Rychlewski L. (2003). 3D-Jury: a simple approach to improve protein structure predictions. Bioinformatics. 19(8):1015-8

Glaser F, Pupko T, Paz I, Bell RE, Bechor-Shental D, Martz E, Ben-Tal N. (2003). ConSurf: identification of functional regions in proteins by surface-mapping of phylogenetic information. Bioinformatics. 19(1):163-4.

Göbel U, Sander C, Schneider R, Valencia A. (1994). Correlated mutations and residue contacts in proteins. Proteins. 18(4):309-317.

Godzik A, Skolnick J. (1992). Sequence-structure matching in globular proteins: application to supersecondary and tertiary structure determination. Proc Natl Acad Sci USA. 89(24):12098-12102.

Godzik A. (2003). Fold recognition methods. Methods Biochem Anal. 44:525-546.

Gracy RW. (1975).Triosephosphate isomerase from human erythrocytes. Methods Enzymol. 41:442-447.

Greene LH, Higman VA (2003). Uncovering network systems within protein structures. J Mol Biol. 334(4):781-91.

Greene LH, Lewis TE, Addou S, Cuff A, Dallman T, Dibley M, Redfern O, Pearl F, Nambudiry R, Reid A, Sillitoe I, Yeats C, Thornton JM, Orengo CA. (2007). The CATH domain structure database: new protocols and classification levels give a more comprehensive resource for exploring evolution. Nucleic Acids Res. 35:D291-D297.

Hardin C, Pogorelov TV, Luthey-Schulten Z. (2002). *Ab initio* protein structure prediction. Curr Opin Struct Biol. 12(2):176-181.

Holm L, Park J. (2000). DaliLite workbench for protein structure comparison. Bioinformatics. 16(6):566-567.

Hooft RW, Vriend G, Sander C, Abola EE. (1996). Errors in protein structures. Nature. 381(6580):272.

Huang W, Petrosino J, Hirsch M, Shenkin PS, Palzkill T. (1996).Amino acid sequence determinants of beta-lactamase structure and activity. J Mol Biol. 258(4):688-703.

Huisman GW, Gray D. (2002). Towards novel processes for the fine-chemical and pharmaceutical industries. Curr Opin Biotechnol. 13(4):352-358.

Jaroszewski L, Rychlewski L, Zhang B, Godzik A. (1998). Fold prediction by a hierarchy of sequence, threading, and modeling methods. Protein Sci. 7(6):1431-40.

Jeffery CJ. (1999). Moonlighting proteins. Trends Biochem Sci. 24(1):8-11.

Jeffery CJ. (2003). Moonlighting proteins: old proteins learning new tricks. Trends Genet. 19(8):415-7.

Jeffery CJ. (2009). Moonlighting proteins – an update. Mol. BioSyst. 5:345-350.

Jeong H, Mason SP, Barabasi AL, Oltvai ZN (2001). Lethality and centrality in protein networks. Nature. 411 (6833), 41-42.

Jeong H, Tombor B, Albert R, Oltvai ZN, Barabasi AL (2000). The large-scale organization of metabolic networks. Nature. 407 (6804), 651-654.

Jones D, Thornton J. (1993). Protein fold recognition. J Comput Aided Mol Des. 7(4):439-456.

Jones DT, Taylor WR, Thornton JM. (1992). A new approach to protein fold recognition. Nature. 358(6381):86-89.

Jones TA, Thirup S.(1986). Using known substructures in protein model building and crystallography. EMBO J. 5(4):819-822.

Kohl L, Callens M, Wierenga RK, Opperdoes FR, Michels PA. (1994).Triose-phosphate isomerase of Leishmania mexicana mexicana. Cloning and characterization of the gene, overexpression in Escherichia coli and analysis of the protein. Eur J Biochem. 220(2):331-338

Kolinski A. (2004). Protein modeling and structure prediction with a reduced representation. Acta Biochim Pol. 51(2):349-71.

Koretke KK, Luthey-Schulten Z, Wolynes PG. (1998). Self-consistently optimized energy functions for protein structure prediction by molecular dynamics. Proc Natl Acad Sci U S A. 95(6):2932-2937.

Koretke KK, Luthey-Schulten Z, Wolynes PG. (2002). Ab initio protein structure prediction. Curr Opin Struct Biol. 12(2):176-181.

Krietsch WK. (1975a). Triosephosphate isomerase from rabbit liver. Methods Enzymol. 41:438-442.

Krietsch WK. (1975b). Triosephosphate isomerase from yeast. Methods Enzymol. 41:434-438.

Krojer T, Garrido-Franco M, Huber R, Ehrmann M, Clausen T. (2002). Crystal structure of DegP (HtrA) reveals a new protease-chaperone machine. Nature. 416(6879):455-459.

Kuhlman B, Baker D (2000) Native protein sequences are close to optimal for their structures. Proc. Natl. Acad. Sci. USA 97:10383-10388.

Kursula I, Partanen S, Lambeir AM, Wierenga RK. (2002).The importance of the conserved Arg191-Asp227 salt bridge of triosephosphate isomerase for folding, stability, and catalysis. FEBS Lett. 518(1-3):39-42.

Kuhlman B, Dantas, G, Ireton GC, Varani G, Stoddard BL, Baker D. (2003). Design of a novel globular protein fold with atomic-level accuracy. Science 302:1364-1368.

Laskowski RA, Luscombe NM, Swindells MB, Thornton JM. (1996). Protein clefts in molecular recognition and function. Protein Sci. 5(12):2438-2452.

Laskowski RA, Moss DS, Thornton JM. (1993). Main-chain bond lengths and bond angles in protein structures. J Mol Biol. 231(4):1049-1067.

Lee J, Liwo A, Scheraga HA. (1999).Energy-based de novo protein folding by conformational space annealing and an off-lattice united-residue force field: application to the 10-55 fragment of staphylococcal protein A and to apo calbindin D9K. Proc Natl Acad Sci USA. 96(5):2025-30.

Lee J, Natarajan M, Nashine VC, Socolich M, Vo T, Russ WP, Benkovic SJ, Ranganathan R. (2008). Surface sites for engineering allosteric control in proteins. Science. 322(5900):438-442.

Lockless SW, Ranganathan R. (1999). Evolutionarily conserved pathways of energetic connectivity in protein families. Science. 286(5438):295-299.

Loeb DD, Swanstrom R, Everitt L, Manchester M, Stamper SE, Hutchison CA 3rd. (1989). Complete mutagenesis of the HIV-1 protease. Nature. 340(6232):397-400.

Luetz S, Giver L, Lalonde J. (2008). Engineered enzymes for chemical production. Biotechnol Bioeng. 101(4):647-653.

Martí-Renom MA, Stuart AC, Fiser A, Sánchez R, Melo F, Sali A. (2000). Comparative protein structure modeling of genes and genomes. Annu Rev Biophys Biomol Struct. 29:291-325.

Martin AC, Orengo CA, Hutchinson EG, Jones S, Karmirantzou M, Laskowski RA, Mitchell JB, Taroni C, Thornton JM. (1998). Protein folds and functions. Structure. 6(7):875-884.

Milenković T, Filippis I, Lappe M, Pržulj N, (2009) Optimized Null Model for Protein Structure Networks. PLoS ONE 4: e5967.

Mirsky AE, Pauling L. (1936). On the Structure of Native, Denatured, and Coagulated Proteins. Proc Natl Acad Sci USA. 22(7):439-447.

Montiel Molina HM, Millan-Pacheco C, Pastor N, del Rio G (2008) Computer-Based Screening of Functional Conformers of Proteins. PLoS Comput Biol. 4(2):e1000009.

Moult J. (2005). A decade of CASP: progress, bottlenecks and prognosis in protein structure prediction. Curr Opin Struct Biol. 15(3):285-289.

Neet KE, Lee JC. (2002).Biophysical characterization of proteins in the post-genomic era of proteomics. Mol Cell Proteomics. 1(6):415-420

Norin M, Sundström M. (2002). Structural proteomics: developments in structure-to-function predictions. Trends Biotechnol. 20(2):79-84.

Olmea O, Rost B, Valencia A. (1999). Effective use of sequence correlation and conservation in fold recognition. J Mol Biol. 293(5):1221-1239.

Olmea O, Valencia A. (1997). Improving contact predictions by the combination of correlated mutations and other sources of sequence information. Fold Des. 2(3):S25-S32.

Ortiz AR, Kolinski A, Rotkiewicz P, Ilkowski B, Skolnick J. (1999). Ab initio folding of proteins using restraints derived from evolutionary information. Proteins. Supplement 3:177-185

Osguthorpe DJ. (2000). Ab initio protein folding. Curr Opin Struct Biol. 10(2):146-152.

Ozkan SB, Wu GA, Chodera JD, Dill KA (2007) Protein folding by zipping and assembly. Proc. Natl. Acad. Sci. USA 104: 11987-11992.

Pazos F, Helmer-Citterich M, Ausiello G, Valencia A. (1997).Correlated mutations contain information about protein-protein interaction. J Mol Biol. 271(4):511-523.

Pearlman DA, Case DA, Caldwell JW, Ross WS, Cheatham TE III, DeBolt S, Ferguson D, Seibel G, Kollman P. (1995) AMBER, a package of computer programs for applying molecular mechanics, normal mode analysis, molecular dynamics and free energy calculations to simulate the structural and energetic properties of molecules. Comp Phys Commun. 91:1-41.

Pearson WR, Lipman DJ.(1988). Improved tools for biological sequence comparison. Proc Natl Acad Sci USA. 85(8):2444-2448.

Pedersen JT, Moult J. (1997a). Ab initio protein folding simulations with genetic algorithms: simulations on the complete sequence of small proteins. Proteins. Supplement 1:179-184.

Pedersen JT, Moult J. (1997b). Protein folding simulations with genetic algorithms and a detailed molecular description. J Mol Biol. 269(2):240-259.

Punta M, Ofran Y. (2008). The rough guide to in silico function prediction, or how to use sequence and structure information to predict protein function. PLoS Comput Biol. 4(10):e1000160.

Radivojac P, Iakoucheva LM, Oldfield CJ, Obradovic Z, Uversky VN, Dunker AK. (2007). Intrinsic disorder and functional proteomics. Biophys J. 92(5):1439-56.

Rennell D, Bouvier SE, Hardy LW, Poteete AR. (1991). Systematic mutation of bacteriophage T4 lysozyme. J Mol Biol. 222(1):67-88.

Rivera MH, López-Munguía A, Soberón X, Saab-Rincón G. (2003). Alpha-amylase from Bacillus licheniformis mutants near to the catalytic site: effects on hydrolytic and transglycosylation activity. Protein Eng. 16(7):505-514.

Robson B. (1999). Beyond proteins. Trends Biotechnol. 17(8):311-315.

Rost B. (1995). TOPITS: threading one-dimensional predictions into three-dimensional structures. Proc Int Conf Intell Syst Mol Biol. 3:314-421.

Rost B. (1999). Twilight zone of protein sequence alignments. Protein Eng. 12(2):85-94.

Russ WP, Lowery DM, Mishra P, Yaffe MB, Ranganathan R. (2005). Natural-like function in artificial WW domains. Nature. 437(7058):579-83.

Rychlewski L, Jaroszewski L, Li W, Godzik A. (2000). Comparison of sequence profiles. Strategies for structural predictions using sequence information. Protein Sci. 9(2):232-41.

Sadowski MI, Jones DT. (2009) The sequence-structure relationship and protein function prediction. Curr. Opin. Struct. Biol. 19:357-362.

Sali A, Potterton L, Yuan F, van Vlijmen H, Karplus M. (1995). Evaluation of comparative protein modeling by MODELLER. Proteins. 23(3):318-326.

Sander C, Schneider R. (1991). Database of homology-derived protein structures and the structural meaning of sequence alignment. Proteins. 9(1):56-68.

Shaw DE, Maragakis P, Lindorff-Larsen K, Piana S, Dror RO, Eastwood MP, Bank JA, Jumper JM, Salmon JK, Shan Y, Wriggers W. (2010) Atomic-level characterization of the structural dynamics of proteins. Science 330: 341-346.

Shehu A, Kavraki LE, Clementi C. (2009) Multiscale characterization of protein conformational ensembles. Proteins 76:837-851.

Simons KT, Bonneau R, Ruczinski I, Baker D. (1999). Ab initio protein structure prediction of CASP III targets using ROSETTA. Proteins. Supplement 3:171-176.

Sippl MJ. (1990). Calculation of conformational ensembles from potentials of mean force. An approach to the knowledge-based prediction of local structures in globular proteins. J Mol Biol. 213(4):859-83.

Skolnick J, Fetrow JS, Kolinski A. (2000). Structural genomics and its importance for gene function analysis. Nat Biotechnol. 2000 Mar;18(3):283-287.

Socolich M, Lockless SW, Russ WP, Lee H, Gardner KH, Ranganathan R. (2005). Evolutionary information for specifying a protein fold. Nature. 437(7058):512-518.

Straathof AJ, Panke S, Schmid A. (2002). The production of fine chemicals by biotransformations. Curr Opin Biotechnol. 13(6):548-556.

Terwilliger TC, Zabin HB, Horvath MP, Sandberg WS, Schlunk PM. (1994). In vivo characterization of mutants of the bacteriophage f1 gene V protein isolated by saturation mutagenesis. J Mol Biol. 236(2):556-71.

Thibert B, Bredesen DE, del Rio G (2005). Improved prediction of critical residues for protein function based on network and phylogenetic analyses. BMC Bioinformatics. 6:213.

Valdar WS. (2002). Scoring residue conservation. Proteins. 48(2):227-241.

Vásquez M. (1996). Modeling side-chain conformation. Curr Opin Struct Biol. 6(2):217-21.

Vendruscolo M, Dokholyan NV, Paci E, Karplus M (2002). Small-world view of the amino acids that play a key role in protein folding. Phys Rev E. 65(6 Pt 1):061910.

Wallace AC, Laskowski RA, Thornton JM. (1996). Derivation of 3D coordinate templates for searching structural databases: application to Ser-His-Asp catalytic triads in the serine proteinases and lipases. Protein Sci. 5(6):1001-1013.

Wallner B, Elofsson A. (2003).Can correct protein models be identified? Protein Sci. 12(5):1073-1086.

Xiang J, Jung JY, Sampson NS. (2004).Entropy effects on protein hinges: the reaction catalyzed by triosephosphate isomerase. Biochemistry. 43(36):11436-11445.

Yarov-Yarovoy V, Schonbrun J, Baker D. (2006).Multipass membrane protein structure prediction using Rosetta. Proteins. 62(4):1010-1025.

Yura K, Yamaguchi A, Go M. (2006) .Coverage of whole proteome by structural genomics observed through protein homology modeling database. J Struct Funct Genomics. 7(2):65-76.

Zaks A. (2001). Industrial biocatalysis. Curr Opin Chem Biol. 5(2):130-6.

Zhang H (2002). Protein Tertiary Structures: Prediction from Amino Acid Sequences. In: ENCYCLOPEDIA OF LIFE SCIENCES. John Wiley & Sons, Ltd: Chichester http://www.els.net/ [doi:10.1038/npg.els.0006101].

Zhang Z, Palzkill T. (2003). Determinants of binding affinity and specificity for the interaction of TEM-1 and SME-1 beta-lactamase with beta-lactamase inhibitory protein. J Biol Chem. 278(46):45706-45712.

5

The Two DUF642 *At5g11420* and *At4g32460*-Encoded Proteins Interact *In Vitro* with the AtPME3 Catalytic Domain

Esther Zúñiga-Sánchez and Alicia Gamboa-de Buen
Universidad Nacional Autónoma de México
México

1. Introduction

The plant cell wall provides structural integrity to plant tissues and regulates cellular growth and form. The cell wall is a dynamic compartment that varies in composition and structure during plant development and in response to different environmental signals. During cell division, the cell plate is rapidly generated. The biogenesis of this new cell wall requires the delivery of vesicles containing newly synthesised material. Cell surface material that includes plasma membrane proteins and cell wall components can be also rapidly delivered to the forming cell plate (Dhonukshe et al., 2006). The three different layers that can compose the cell wall are the middle lamella, primary cell wall and secondary cell wall. The middle lamella, which is a pectinaceous interface, is deposited soon after mitosis to create a boundary between the two daughter nuclei and is important for the adhesion of neighbouring cells. The primary cell wall is deposited throughout cell growth and expansion. These two processes require a continuous synthesis and exportation of cell wall components that have to be reorganised in the cell wall network. The secondary cell wall is deposited when cell growth has ceased and is not present in all cell types.

1.1 Polysaccharide composition of the cell wall

The primary cell wall is composed of diverse polysaccharides (85-95%) and cell wall proteins with different functions (5-15%, CWP). Cellulose, hemicelluloses (e.g., xyloglucans) and pectins (e.g., homogalacturonans) are the main types of polysaccharides present in cell wall. Cellulose microfibrils confer rigidity to the cell wall and interact with hemicelluloses to provide structure to the network. These polysaccharide interactions could restrict access of enzymes to their substrates; however, this network can be modified during plant development by different proteins that interact with the network components or by enzymes that modify the polysaccharides (Harpster et al., 2002). The polysaccharides are not the only contributor to cell wall integrity during plant development. Recently, it was demonstrated that the presence of cellulose is essential to maintain the polar distribution of proteins at the plasma membrane. The polar distribution of PIN transporters for the phytohormone auxin is disrupted by a pharmacological interference with cellulose or by mechanical interference with the cell wall (Feraru et al., 2011). Pectins, which are a major component of primary cell wall, are a large group of complex polysaccharides that are synthesised in the Golgi and transported to the cell wall by secretory vesicles (Sterling et al., 2001). Methylesterification of homogalacturonan (HG) occurs in the plant Golgi apparatus, possibly by a S-adenosylmethionine (SAM) methyltransferase named Cotton Golgi-Related-3 (CGR3) (Held et al., 2011). HG is delivered to the cell wall in a highly methylesterified state, and the modulation of this state is a very important process in plant development. Highly esterified pectins are present in the proliferating zone of different tissues, whereas the cell walls of differentiating cells present abundant non-esterified pectins (Barany et al., 2010).

1.2 Protein composition of the cell wall

The cell wall composition is continuously modified by enzyme action during growth and development and in response to environmental conditions (Cassab, 1998). Proteins with enzyme activity and modulatory activity are present with different abundances in different cell types. Approximately 400 cell wall proteins that have been detected in cell wall proteomes have been classified into eight categories on the basis of predicted biochemical functions (Jamet et al., 2006). Members of seven of the eight groups have been previously defined as cell wall proteins involved in different aspects of cell wall dynamics. Many proteins have been detected in cell wall proteomes isolated from apoplastic fluids obtained from seedlings and rosette leaves (Charmont et al., 2005; Boudart et al., 2005), vegetative tissue that included etiolated hypocotyls and stem (Ishrad et al., 2008; Minic et al., 2007) and cell suspension cultures (Chivasa et al., 2002; Bayer et al., 2006, Bordereis et al., 2002). These proteins present a domain with an unknown function and are grouped together. The study of the function of the different families of this group of proteins will provide information about the dynamic processes of the cell wall.

1.2.1 Proteins acting on polysaccharides

Xyloglucan endotransglycosylase/hydrolase (XTH) is a family of glycosyl hydrolases that transglycosylate xyloglucan to allow expansive cell growth. These hydrolases are involved in cell growth, fruit ripening, and reserve mobilisation following germination in xyloglucan-storing seeds. In *Arabidopsis*, 33 genes have been identified that code for these hydrolases. Different temporal and spatial expression patterns for these *XTH* genes suggest that this family is involved in the change of cell wall properties related to every developmental stage. For example, *XHT5* is expressed in hypocotyls, root tips, and anther filaments, whereas *XHT24* is localised in vasculature tissue from the cotyledons, leaves, and petals. However, there is also an overlapping of the *XTH* gene expression pattern that suggests a combinatorial action of this enzyme group (Becnel et al., 2006).

Pectin modification is catalysed by a large family of pectin methylesterases (PMEs). In *Arabidopsis*, 66 genes have been suggested to potentially encode PMEs and are expressed differentially during organ and tissue development. A pro-domain is present in approximately 70% of the *Arabidopsis* PME family members (Micheli, 2001). It has been suggested that this domain has an inhibitory function during transportation to the cell wall by vesicles. The carboxylic fragment with the catalytic domain has been detected in cell wall proteomes, but the complete protein is required for secretion (Wolf et al., 2009). The interaction of PME with proteins that inhibit its activity, which are called pectin methyesterase inhibitors (PMEIs), contributes to the modulation of the degree of the methylesterified state of the pectin in the cell wall during different developmental processes (Pelloux et al., 2007). During pollen germination, the pollen tube wall presents highly methylesterified pectins in the tip region and weakly methylesterified pectins along the tube. It has been suggested that the activity of PMEs during pollen tube growth is highly regulated by PMEIs (Dardelle et al., 2010). Local relaxation of the transmitting tract cell wall also results from changes in the methylesterification of pectins that possibly facilitate the growth of the pollen tubes in the extracellular matrix of this female tissue (Lehner et al., 2010). An important role of pectin modifications in the regulation of cell wall mechanics in the apical meristem tissue has also been suggested (Peaucelle et al., 2011). The demethylesterification of pectin by PME activity results in random and contiguous patterns of free carboxylic residues. These contiguous patterns promote Ca^{++} binding, which generates a rigid cell wall. PMEs might also be involved in maintaining apoplastic Ca^{++} homeostasis. PME activity has been suggested to maintain apoplastic Ca^{++} homeostasis during heat shock. The resulting cell wall remodelling maintains the plasma membrane integrity to confer thermotolerance to the soybean (Wu et al., 2010). The random release of protons promotes pectin degradation by polygalacturonases, which are enzymes that also affect the pectin network. Polygalacturonases (PGs) promote pectin disassembly and might be responsible for various cell separation processes. PG activities are associated with seed germination, organ abscission, anther dehiscence, pollen grain maturation, fruit softening and decay, and pollen tube growth. In *Arabidopsis*, 69 genes encode PGs with different spatial and temporal patterns. For example, *At1g80170* is specifically expressed in the anther and pollen (González-Carranza et al., 2007).

Expansins are cell wall proteins that modify the mechanical properties of the cells to enable turgor-driven cell enlargement. Expansin genes are highly conserved in higher plants, and there are four different expansin families in plants. Multiple expansin genes are often expressed in association with developmental events such as root hair initiation or fruit growth. They are also involved in processes such as fruit ripening and abscission, although cell wall modification occurs without expansion. Expansins may also be involved in embryo growth and endosperm weakening during germination (Sampedro and Cosgrave, 2005). The localised expression of expansins is associated with the meristems and growth zones of the root and stems (Reinhardt et al., 1998).

1.2.2 Oxido-reductases

Peroxidases are implicated in many physiological phenomena that include cross-linking of cell wall components, defence against pathogens, and cell elongation. These enzymes have a great variety of substrates and can regulate growth by controlling the availability of elongation-promoting H_2O_2 in the cell wall (Passardi et al., 2004). In *Arabidopsis*, 73 genes have been reported to code for putative peroxidases (Valério et al., 2004), and AtPrx33 and AtPrx34 function is specifically related to root elongation (Passardi et al., 2006).

Germins are oligomeric enzymes with oxalate oxidase activity that are associated with the extracellular matrix. In *Arabidopsis*, this family contains 12 members that are expressed in almost every organ and developmental stage. *AtGer1* has been implicated in germination, whereas *AtGer2* is involved in seed maturation (Membré et al., 2000).

1.2.3 Proteases

Proteases cleave peptide bonds and are classified into four catalytic classes: Cys proteases, Ser carboxypeptidases, metalloproteases and Asp proteases. The *Arabidopsis* genome encodes 826 proteases that are classified into 60 families with high functional diversity. Plant proteases are key regulators of different biochemical processes that are related to meiosis, gametophyte survival, embryogenesis, seed coat formation, cuticle deposition, epidermal cell fate, stomata development, chloroplast biogenesis, and local and systemic defence responses (van der Hoorn, 2008). Some proteases have been detected in cell wall proteomes, especially in cell suspension cultures.

1.2.4 Proteins that have interacting domains with no enzymatic activity

LRR proteins are frequently implicated in protein-protein interactions and are localised in the different subcellular compartments (Kajava, 1998). The LRR superfamily includes polygalacturonase-inhibiting proteins (PGIPs) that are present in the cell wall and are involved in disease resistance as well as growth and development (Di et al., 2006). FLOR 1, a putative PGIP protein, has been detected in cell wall proteomes but is also localised intracellularly, as more than 70% of the PGIP in *Pisum sativum* was reported to be distributed in the cytoplasm (Acevedo et al., 2004; Hoffman & Turner, 1984).

Pectin methyl esterases inhibitors (PMEIs) are a diverse group of proteins that belong to the family of invertase inhibitors (INHs). PMEIs share with INHs a domain that is characterised by four conserved cysteine residues that can form two disulfide bonds (Juge, 2006). In *Arabidopsis*, there is an spatial patterning of cell wall PMEI at the pollen tip (Röckel et al., 2008).

Lectins are a diverse group of carbohydrate specific binding proteins that are involved in signal transduction (Lannoo et al., 2007). This group of proteins has interacting domains but does not show catalytic activity. The group presents with varying cellular localisation, which suggests a role in signal transduction between the different cellular compartments (Van Damme et al., 2004).

1.2.5 Proteins involved in signalling

In plants, there is a large subclass of receptor-like kinases that have extracellular LRRs in the receptor domain and are involved in signal transduction during development or defence (Clark et al., 1997). Arabinogalactan proteins (AGPs) are hydroxyproline-rich glycoproteins that are also involved in signalling. This family contributes to defensive, adhesive, nutrient and guidance function during pollen-pistil interactions (Cassab, 1998).

1.2.6 Proteins related to lipid metabolism

Lipases (LTPs) are hydrolytic enzymes with multifunctional properties. GDSL lipases are mainly involved in the regulation of plant development, morphogenesis, synthesis of secondary metabolites and defence responses (Ruppert et al., 2005).

1.2.7 Structural proteins

LRR-extensins were the only group of structural proteins detected in cell wall proteomes. This family may be involved in the local regulation of cell wall expansion. Eleven genes have been described in *Arabidopsis*; four of them are pollen specific (Baumberger et al., 2003).

1.2.8 Unknown proteins

Approximately 5 to 30% of the total proteins from different cell wall proteomes have been classified as hypothetical, expressed, putative, unknown or with a domain of unknown function (DUF), especially in cell suspension culture. A domain is considered to be a discrete portion of a protein that folds independently of the rest of the protein and possesses its own function. Eight DUF protein families (DUF26, DUF231, DUF246, DUF248, DUF288, DUF642, DUF1005, DUF1680) are represented by one (or more) member(s) of the cell wall proteomes.

DUF26 is a plant-specific protein family composed of 40 members in *Arabidopsis*. Some members include DUF26 receptor-like kinases (RLKs), which are also known as cysteine-rich RLK (CRKs). These proteins are involved in pathogen resistance and are transcriptionally induced by oxidative stress and pathogen attack (Wraczeck et al., 2010). *At5g43980* encodes a protein present in the apoplastic fluid from rosette leaves that has been described as a plasmodesmal protein (PDLP1) involved in cell-to-cell communication processes (Thomas et al., 2008). The other DUF26 protein, which was detected in the cell wall proteome from cell suspension cultures, has not yet been assigned a function.

DUF231 is present in the proteins of the *TRICHOME BIREFRINGENCE/TRICHOME BIREFRINGENCE-LIKE* (TBR/TBL) plant family with 46 members in *Arabidopsis*. The role of this family in cellulose biosynthesis has been recently described; *tbr* mutants presented decreased levels of crystalline secondary wall cellulose in trichomes and stems (Bischoff et al., 2010a). Loss of TBR also results in increased PME activity and reduced pectin esterification, which suggests that TBL/DUF231 proteins are "bridging" proteins that crosslink different cell wall networks (Bischoff et al., 2010b). *At5g06230* (TBL9) was found in a cell wall proteomic analysis of etiolated hypocotyls (Ishrad et al., 2008).

The domain unknown function 246 is considered to be a GDP-fucose o-fucosyltransferase domain in animals. This protein family has 16 members in *Arabidopsis,* and one of them, *At1g51630*, was detected in the proteome of cell suspension cultures.

DUF248 is a putative methyltransferase-related family of proteins with an ankyrin-like protein domain that is related to dehydration-responsive proteins. There are 29 proteins of this family in *Arabidopsis*, but only one, *At5g14430*, has been described in the cell wall proteome of cell suspension cultures (Bayer et al., 2006).

DUF288 is not a plant-specific family; this domain is also found in *Caenorhabditis elegans* proteins. In *Arabidopsis,* there are two members: *At2g41770* and *At3g57420*. *At3g57420* encode protein was purified from the apoplastic fluid of the cell wall proteome of rosette leaves (Boudart et al., 2005).

The DUF1005 domain has five integrants in *Arabidopsis* with two members that are similar to IMP dehydrogenase/GMP reductase from *Medicago trunculata*. The integrant isolated from the cell wall proteome of mature stems (*At4g29310*) does not have the other domain (Minic et al., 2007).

Two loci are described in *Arabidopsis* for the DUF1680 family, and one of them was purified from the cell wall proteome of mature stems.

The most important family of unknown proteins detected in cell wall proteomes is DUF642, which is a highly conserved plant-specific family that is present in angiosperms and gymnosperms (Albert et al., 2005, Vázquez-Lobo, personal communication). *Arabidopsis* has ten members. The *At3g08030*-encoded protein is present in all cell wall proteomes and is the only unknown protein that was also detected in a seed proteome from the *Arabidopsis* accession Cape Verde Island (Cvi) that has deeper seed dormancy (Chibani et al., 2006). *At2g41800* and *At1g80240*-encoded proteins were only found in cell suspension cultures (Bayer et al., 2006), whereas *At5g25460*-encoded protein was found in vegetative and cell wall suspension cultures. *At4g32460* and *At5g11420*-encoded proteins were both detected in apoplastic and vegetative tissues. The consistent presence of 6 members of this family in all cell wall proteomes suggest that the biochemical function of the DUF642 family is related to the regulation of the activity of cell-wall-modifying enzymes at different stages of plant development.

1.3 DUF642 family

The DUF642 protein family is highly conserved, is widespread in plants, and might be involved in important basic developmental processes. Members of this family have been observed in basal angiosperms such as *Amborella*, in both monocots and dicots and also in gymnosperm species. The relevance of the DUF642 family to plant evolution was discussed by Albert and collaborators (2005). The proteins encoded by the DUF642 gene family have a unique, highly conserved domain with no assigned function that shares similarity with the galactose-binding domain. The ten members of this family identified in *Arabidopsis* contain a signal peptide of 20 to 30 amino acids in the N-terminus region that could promote their localisation in the endomembrane system or in the cell wall. Three of the ten *Arabidopsis* genes (*At1g29980*, *At2g34510* and *At5g14150*) encode proteins have been described as glycosil-phosphatidyl-inositol anchored proteins (Figure 1) (Borner et al., 2003, Dunkley et al., 2006). The *At2g41800*-encoded protein has been detected in the *Arabidopsis* cell wall proteome. The proteins encoded by *At5g11420* and *At2g34510* contain a ATP/GTP binding site motif that has been described in many proteins involved in signal transduction processes.

Although a function has not yet been assigned for this family, it has been suggested that some members could be involved in different developmental processes. Organ-specific expression has been described for the flowers of two DUF642 members, *At3g08030* and *At5g11420* (Wellmer et al., 2004), and for the stems for a DUF642 *Medicago sativa* gene (Abrahams et al., 1995). *At4g32460*, *At5g14150* and *At2g41800* have been described as papillar cell-specific genes in flowers (Tung et al., 2005). Changes in DUF642 gene expression have been also detected under specific environmental conditions. Saline stress promotes the expression of *At2g41810* (Kreps et al., 2002), and an RNA increase in the three DUF642 *Arabidopsis* homologs (*At3g08030*, *At5g25460* and *At4g32460*) was described during the priming and germination of *Brassica oleracea* seeds (Soeda et al., 2005).

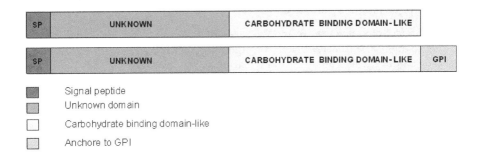

Fig. 1. DUF642 proteins have a basic structure divided into two subdomains and a signal peptide. N-terminus subdomain has not function or putative function assigned while C-terminus subdomain has homology with a carbohydrate binding domain. Some DUF642 proteins present in their C-terminus a GPI anchored motive.

We characterised the plant-specific DUF642 protein family using different approaches. We determined mRNA expression in different plant tissues, characterised sequence features and detected the potential interaction of proteins with two members of this family in *Arabidopsis* (*At5g11420* and *At4g32460*-encoded proteins). The proteins identified by LC/MS/MS analysis were the leucine-rich repeat protein FLOR1 (FLR1), a vegetative storage protein (VSP1), and a ubiquitous pectin methylesterase isoform (PME3) isolated from *Arabidopsis* flowers and leaves. Based on the structural characteristics of the DUF642 family of proteins and the associated affinity chromatography analyses, we propose that these proteins could interact specifically with other cellular components via their DUF642 domain and are therefore potentially involved in developmental plant processes. Our results provide a starting point for defining the function of the DUF642 family in plant development.

2. Materials and methods

2.1 Plant material and sample collection

Arabidopsis thaliana from the Columbia (Col) ecotype plants were grown on MS plates (1X Murashige and Skoog basal salt mixture, 0.05% MES, 1% sucrose as carbon source and 0.8% agar) in a REVCO growth chamber under a long photoperiod (16-h light 8-h darkness) at 20°C. Fifteen-day-old seedlings were transferred to pots containing Metro-Mix 200 (Scotts Company) soil and grown under the same controlled conditions.

2.2 Reverse transcriptase–polymerase chain reaction (RT-PCR)

Arabidopsis samples from different tissues were collected from 15-day-old seedlings and flowering plants, immediately frozen in liquid nitrogen and stored at -80°C until analysis. Total RNA from different tissues was isolated using TRIZOL according to the supplier's instructions (INVITROGEN™). cDNA templates for the amplification by PCR were prepared using SuperScript II reverse transcriptase (INVITROGEN™) according to the manufacturer's instructions. Based on the sequence of each gene member of the DUF642 family of *Arabidopsis*, the following primers were synthesised:

At2g41800: F 5'tcctcctcctatctctctgc 3' and R 5'aaacggttctcttcctgc 3';
At2g41810: F 5'atgggccaaaaaaacac 3' and R 5'atgtctctcgttctctctc 3';
At3g08030: F 5'ggttcccaaagccattattc 3' and R 5'acaatctcgtcaatgacagg3';
At5g25460: F 5'cttccttcttttcatcgcc 3' and R 5'acgagaaatcatcgctcc 3';
At5g11420: F 5'ccatgggcttcagtgacgggatg 3' and R 5'agatctgagtgtcttttcccgc 3';
At4g32460: F 5'gtgatagtgcttcttctccttcac 3' and R 5' agcgacgaatctcaatgac 3';
At1g80240: F 5'aaaagcagcactcctcttag 3' and R 5' atcattggtccctcacaac 3';
At1g29980: F 5'ccgagcaacaatagatgc 3' and R 5'actgtagaacgcaactctgg 3';
At2g34510: F 5'ttggtctctccattgtggc 3' and R 5'ccttaacgtcatcaatcacagg 3';
At5g14150: F 5´ttgcgcctcttcagatttt3'and R 5'cttctcaccagagccagtcc 3'.

Polymerase chain reaction (PCR) was performed under the following conditions: 94°C 5 min; 35 cycles of 94°C 30 sec, 60-62°C 30 sec, 72°C 1 min 30 sec, 72°C 5 min.

2.3 Sequence analysis and database search

The 10 DUF642 protein sequences of *Arabidopsis* were obtained from GenBank (NP_973938: *At1g29980*; NP_178141: *At1g80240*; AAC02768: *At2g41800*; AAC02767, NP_181712: *At2g41810*; AAC26689: *At2g34510*; AAO00904: *At3g08030*; ABF19001: *At4g32460*; NP_196919: *At5g14150*; AAN31807: *At5g11420* and AAP37805: *At5g25460*). A multiple sequence alignment, using only the DUF642 protein domain, was performed using ClustalW from the Bio Edit Sequence Alignment Editor. The possible secondary structure of the proteins coded for by *At5g11420* and *At4g32460* was compared on-line using the Draw an HCA (Hydrophobic Cluster Analysis) program (http://ca.expasy.org/tools/) as described in Gaboriaud et al. (1987).

2.4 Recombinant 5xHis-tagged DUF642 proteins and the resin-bound DUF642 protein affinity column

The entire open reading frame of the DUF642 genes *At5g11420* and *At4g32460*, without the signal-peptide-coding region, was amplified using PCR. The primers used for the *At5g11420* were MET11420 (5'ccatgggcttcagtgacgggatg3'), which includes an in-frame ATG, and primer 11420FIN2 (5'agatctagtgtcttttcccgca3'). For the amplification of the carboxyl-terminus truncated protein, the *At5g11420* (Δ11420) forward primer MET11420 and the reverse primer 11420FIN3 (5´agatctcggcttacgagcactgag3´) were used. *At4g32460* was amplified using the following primers: MET32460 (5'ccatgggcttcaatgatggactactacc3') and 32460FIN2 (5'agatctgcgtaaaacgtactgtaga3'). The amplified regions of these genes were cloned into the pQE60 vector using the NcoI and BglII restriction sites. A negative control was performed using the empty pQE60 vector. Protein expression and purification were performed following the supplier's instructions, and the recombinant proteins with the histidine tail were detected using western blot analysis with a Ni-NTA conjugate (QIAGEN). The three recombinant proteins were eluted as a single band and were identified to have the histidine tail. No protein was detected when the empty vector was used. The elution process was the only step omitted when the column was prepared for each recombinant protein.

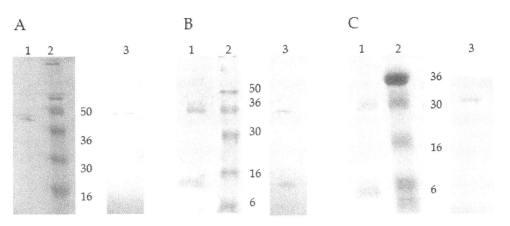

Fig. 2. Recombinant 5xHis-tagged DUF642 proteins.
A) Purification of the 32460 recombinant protein. 12% PAGE Gels were stained with Coomassie Blue. The column was eluted with 250 mM Imidazole (Lane 1). Western Blot of the eluted fraction (NiNta beads with phosphatase alkaline secondary antibody). The band of approximately 40 kDa corresponds to the calculated molecular weight for this protein (Lane 3).
B) Purification of the 11420 recombinant protein. 12% PAGE Gels were stained with Coomassie Blue. The column was eluted with 250 mM Imidazole (Lane 1). Western Blot of the eluted fraction (NiNta beads with phosphatase alkaline second antibody). The band of approximately 40 kDa corresponds to the calculated molecular weight for this protein (Lane 3).
C) Purification of the Δ11420 recombinant protein. 12% PAGE Gels were stained with Coomassie Blue. The column was eluted with 250 mM Imidazole (Lane 1). Western Blot of the eluted fraction (NiNta beads with phosphatase alkaline second antibody). The band of approximately 32 kDa corresponds to the calculated molecular weight for this protein (Lane 3).

2.5 Affinity chromatography of flower or leaf protein extracts

Frozen flowers or leaves from *Arabidopsis* plants (10-20 g) were ground with a mortar and pestle and placed in two 40 ml tubes with 14 ml of extraction buffer (50 mM Tris-HCl pH 7.5, 3 mM MgCl$_2$, 1 mM PMSF). The crude homogenate was centrifuged at 15,000 xg for 30 min, and in the case of DUF642 affinity columns, the supernatant was loaded onto a previously equilibrated DEAE-Sephacel column (2x 10 cm) with extraction buffer at 4°C. The resulting fraction was then used for affinity chromatography. The affinity column was prepared beforehand as described above and equilibrated with extraction buffer. The protein extracts from the different tissues were mixed for 1 h with the prepared resin at 25°C using gentle agitation in a ratio of 10 ml of extract/0.2 ml of agarose. The column was washed with 50 mM Tris-HCl pH 7.5, 5 mM MgCl$_2$ (50 vol) buffer to remove unbound proteins. Bound proteins were eluted with the same buffer containing different NaCl concentrations (100 to 1000 mM). These fractions were precipitated with cold acetone. Agarose and the empty vector column were used as negatives controls, and no bound proteins were detected (Gamboa et al., 2001).

The fractions obtained in the affinity chromatography assays were analysed on denaturing 12% SDS-PAGE gels and stained with silver. Bands of interest were extracted from the gels and sent to the Proteomics Platform of the Eastern Genomics Center, Quebec, Canada, where the in-gel digest and mass spectrometry experiments were performed. Tryptic digestion was performed according to Shevchenko et al. (1996) and Havlis et al. (2003). Peptide samples were separated by online reversed-phase (RP) nanoscale capillary liquid chromatography (nano/LC) and analysed by electrospray mass spectrometry (ES/MS/MS).

Database searching. All MS/MS samples were analysed using Mascot (Matrix Science, London, UK; version 2.2.0)

Criteria for protein identification. Scaffold (version Scaffold-01_07_00, proteome Software Inc. Pórtland Oregon, OR) was used to validate MS/MS-based peptide and protein identifications. Peptide identifications were accepted if they could be established at greater than 95.0% probability as specified by the Peptide Prophet algorithm (Keller et al., 2002). Protein identifications were accepted if they could be established at greater than 95.0% probability and contained at least 2 identified peptides. Protein probabilities were assigned by the Protein Prophet algorithm.

Only one protein was identified for the protein bands derived from the two chromatography steps, DEAE-Sephacel and affinity chromatography (11420 and 32460 column affinity protocols).

3. Results and discussion

3.1 Gene structure of the DUF642 family in *Arabidopsis thaliana*

The DUF642 domain was only present in the ten *Arabidopsis* members described before, and all members had the same gene structure, which consisted of three exons and two introns (Figure 3). The first intron encoded the signal peptide, and an alternative usage of the first exon was detected for *At1g29980* and *At3g08030*. The first intron was also included in the mRNA sequence for the *At3g08030* gene. The expression of two different mRNAs has been found in different tissues, which suggests a possibly different protein subcellular localisation.

3.2 DUF642 members are widely expressed in all *Arabidopsis thaliana* plant tissues

The RT-PCR expression analysis of the ten DUF642 genes in different tissues including seedlings, stems, cauline leaves, rosette leaves, flowers, inflorescences and roots is shown in Figure 4. The genes with broad expression patterns are *At1g80240*, *At5g11420*, *At5g25460* and *At2g41800*, whereas *At1g29980* and *At4g32460* were not detected in cauline leaves. *At2g41810* expression was restricted to inflorescence tissue. The *At2g41810*-encoded protein exhibits 81% identity and 89% similarity to the *At2g41800*-encoded protein. In the inflorescence tissue, the *At2g41800* transcript contained an additional region of 100 bp corresponding to the first intron, which suggests an alternative use of the first exon described for *At3g08030* and *At1g29980*. The gene with the most divergent sequence in the family, *At5g14150*, was also detected in the stem, flower, inflorescence, and root tissues and was detected at low levels in cauline leaves.

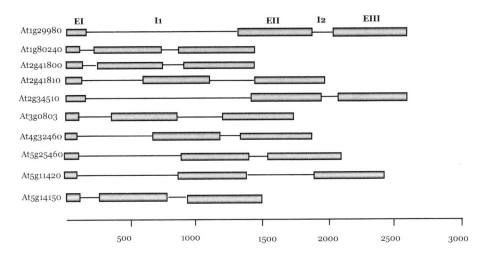

Fig. 3. Gene structure of DUF642 family in *Arabidopsis thaliana*.
EI: Exon 1, **I1**: Intron 1, **EII**: Exon 2, **I2**: Intron 2, **EIII**: Exon3

Our results are consistent with the microarray data described in the Gene Investigator Atlas (http://www.genevestigator.ethz.ch/), except for the *At2g41810* gene. We did not find *At2g41810* expression in the roots, but the Atlas indicated high expression. However, Kreps and collaborators (2002) demonstrated that the expression of this gene in the roots is induced by NaCl stress. These discrepancies in the results obtained in different studies could therefore be related to the different growth conditions used. Spatio-temporal expression analyses of this family will provide important information about its function. Cell-type-specific expression in the roots of the auxin-inducible DUF642 genes *At2g41800* and *At4g32460* was recently reported (Goda et al., 2004; Salazar-Iribe & Gamboa-deBuen, unpublished data).

Transcriptomic analyses suggest that the expression of this family of genes is also affected by different environmental conditions. The expression of genes that encode DUF642 proteins could be inhibited or stimulated by different pathogens. Indeed, invasion by necrotrophic pathogens or insect attack has been shown to significantly reduce the expression of *At5g11420*, *At5g25460*, *At4g32460* and *At1g29980* in plant tissues (Hu et al., 2008; Ehlting et al., 2008). Conversely, an increase of DUF642 gene expression in response to biotrophic organisms has been reported in *Arabidopsis* transcriptomic analyses of sink-heterologous structures, such as galls. Furthermore, the *At3g08030* and *At1g29980* genes have been found to be up-regulated in response to *Agrobacterium tumefaciens* and *Rhodococcus fascians* invasion (Depuydt et al., 2009, Lee et al., 2009). *At1g29980* has also been shown to be highly expressed in the giant cells induced by the root-knot nematode, *Meloidogyne incognita* (Barcalá et al., 2010), and the development of such sink structures is related to an increase in auxin (Grunewald et al., 2009). The study of the effect of nematode invasion on the gene expression of the DUF642 family will provide important functional insights.

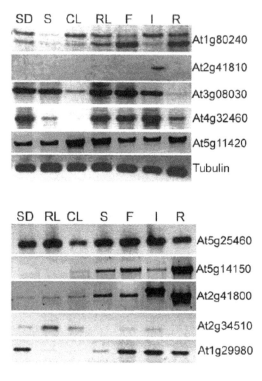

Fig. 4. RT-PCR expression of *Arabidopsis thaliana* DUF642 genes in various tissues. Seedlings (SD), rosette leaves (RL), cauline leaves (CL), stems (S), flowers (F), inflorescences (I), and roots (R). The expression of tubulin was analyzed simultaneously as an internal standard.

3.3 Comparison of the primary sequence of the ten *Arabidopsis thaliana* DUF642 family members

The DUF642 gene family encodes proteins with an estimated molecular mass ranging from 39 to 44 kDa. These proteins contain the DUF642 amino acid domain, preceded by a 20-30 amino acid signaling peptide on the amino terminus. This signaling peptide could be involved in the cell wall localisation of DUF642 proteins in several plant organs. Alignment analysis of the ten *Arabidopsis* members shows an extensive conservation of the DUF642 domain; the percentage of identical and similar amino acids varies from 30% to 85% and 43% to 92%, respectively (Figure 5A). About 30% of the amino acids distributed throughout the sequence of the DUF642 domain are hydrophobic. These residues are not identical, but they are similar among the different proteins. The comparison of the hypothetical secondary structure of *At5g11420* and *At4g32460*-encoded proteins shows that the hydrophobic clusters present are similar (Figure 5B). Four conserved cysteine residues are present in all of the sequences as previously described for the pectin methyl esterase inhibitors localised in the cell wall (Juge, 2006). Because no catalytic activity has yet been assigned to the DUF642 domain, this family could be involved in specific carbohydrate or protein interactions.

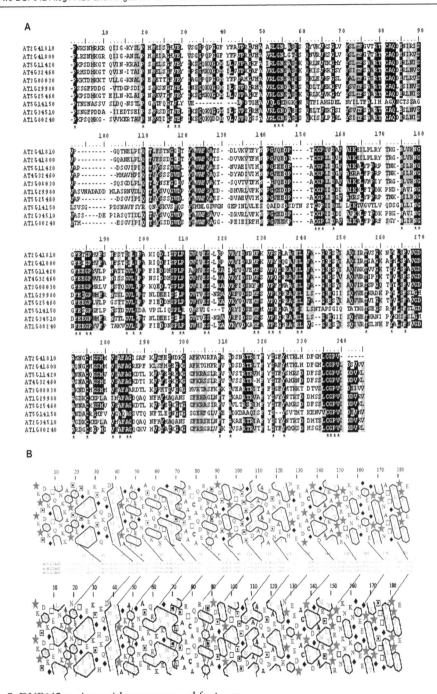

Fig. 5. DUF642 amino acid sequence and features.
(A) Clustal W alignment (BioEdit) of the DUF642 domain of the 10 *Arabidopsis* proteins is

shown. The N-terminal region (comprising the signal peptide) was eliminated for the alignment. Shading indicates conserved amino acid and dark shading indicates identities. (B). Secondary structure comparison of 11420, top, and 32460, bottom. The initial 180 amino acid sequences of the DUF642 domain of both proteins are compared using "Draw an HCA" online program (http://ca.expasy.org/tools/) (Gaboriaud et al., 1987). Amino acids forming putative hydrophobic clusters are grouped together. Compare similar patterns in both sequences. Star: P; dotted square: S; rhomb: G, and empty square: Y residues; other amino acids in standard abbreviation.

Most of the members of the DUF642 family have a broad expression pattern in different plant tissues. A putative redundancy of function in this family should be considered because of the high conservation of the DUF642 domain; however, it is important to describe the organ, cell type and specific stress-related expression patterns for each gene to determine the individual gene function (Wellmer et al., 2004).

3.4 DUF642 proteins have specific interactors in the flowers and leaves of *Arabidopsis thaliana*

Recombinant 32460 protein interacts *in vitro* with the LRR protein FLR1 (Q9LH52, *At3g12145*), with VSP1 (Q93VJ6, *At5g24780*) and with PME (Q9LUL7, *At3g14310*) in flowers, whereas in leaves, it interacts with the same PME (*At3g14310*) (Figure 6). The recombinant 32460 protein interacts *in vitro* with three proteins with sizes of 38 kDa, 37 kDa and 29 kDa from the flowers (Figure 6A). These proteins were identified as FLR1, PME, and VSP1, respectively (Figures 8A, B and C). It is important to note that FLR1 was not eluted by 500 mM NaCl, and VSP1 is only present in this fraction as determined in the interaction assay using the *At5g11420*-encoded protein. A 37 kDa band was purified in the three salt fractions from leaf extracts and was identified as the same PME isoform described for the flowers. A 29 kDa band was also eluted, and this protein was identified as a possible auxin-binding protein (Figure 6B). For all protein bands analysed, only a significant hit was assigned, as described in the material and methods.

The recombinant DUF642 11420-protein interacts *in vitro* with FLOR1 and VSP1 in flowers, but in leaves, it only interacts with PME (Figure 7). A high-purity protein fraction with two bands was obtained from the 11420-affinity column after the floral crude protein extracts were purified over several steps (Figure 7B). Different ionic strengths were used during elution; one 38 kDa band was eluted at 100 and 200 mM NaCl, whereas a 29 kDa band was obtained at 200 and 500 mM NaCl (see arrows in Figure 7B). The 38 kDa protein was identified as FLR1 (12% coverage) and the 29 kDa band as VSP1 (11% coverage), as described in the methods (Figures 8A and B). A Δ11420 protein without the carboxylic terminus that included the most divergent amino acid sequence was also used as a ligand. FLR1 was the only purified protein, which suggests that the carboxylic region is important for interaction with VSP1 (Figure 7C). *At5g11420* is expressed in all *Arabidopsis* tissues, and therefore, we were interested in the determination of the proteins in the leaves that interact with the *At5g11420*-encoded protein. The same procedure, using the affinity column with a leaf extract protein fraction, was used. In the first two fractions, two bands of 45 kDa and 32 kDa were detected. In the 500 mM NaCl fraction, three major bands of the following sizes were detected: 45 kDa, 32 kDa and 14 kDa (Figure 7D). The identified 32 and 14 kDa bands correspond to a PME (40% coverage, Q9LUL7, *At3g14310*). The PME 14 kDa band was also identified when the Δ11420 protein was used as the ligand (Figure 7E). The two lower-molecular-weight bands contained the carboxyl region that includes the catalytic domain of the PME, and therefore, it is possible that the differences in their electrophoretic mobility are the result of post-translational modifications.

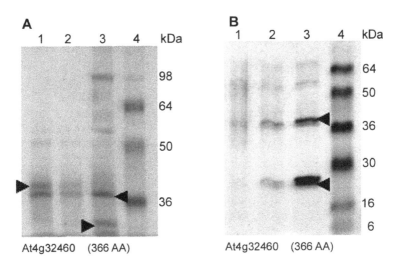

Fig. 6. 32460-protein *in vitro* interactors.
(A) Recombinant 32460 amino acid sequence. (B C) affinity chromatography assays of 32460 interactors from DEAE-Sephacel flow-through protein fraction from *Arabidopsis thaliana* flowers (B) and leaves(C). Silver staining of 12% SDS-PAGE gel showing: (1) NaCl 100 mM, (2) NaCl 200 mM, (3) NaCl 500 mM elution fractions, and (4) molecular weight reference. (B) Flower interactors of the 32460 recombinant protein. In (1) and (2) two main protein bands are seen; with molecular masses of 38 and 37, corresponding to FLR1 (arrow in (1)) and PME (upper arrow in (3)) respectively. In (3) the two bands with molecular masses of 37 and 29 were identified as PME and VSP1 respectively (see arrows in (3)). (C) Leaf interactors of the 32460 protein. Fraction (3) was highly enriched with two bands with molecular masses of approximately 37 and 29. The 37 kDa band was identified as the catalytic domain of a PME, while the 29 kDa band was identified as a possible auxin-binding protein.

The proteins that interacted *in vitro* with the DUF642 11420 and 32460 proteins, i.e., FLOR1 and AtPME3, were detected in the cell wall proteomes (Figures 7A, B and C). Similar expression patterns reflect a possible *in vivo* interaction. FLOR1 is an LRR protein related to polygalacturonase inhibitors (PGIPs) that are highly expressed in vascular and meristem tissues. An intracellular localisation of FLOR1 has been also reported (Acevedo et al., 2004). AtPME3 (*At3g14310*) is expressed in the vascular tissue of seedlings, leaves, stems and roots and is involved in adventitious root formation (Guénin et al., 2011). Recently, we demonstrated that *At4g32460* is also expressed in the meristems and in vascular tissue (Zúñiga-Sánchez & Gamboa-deBuen, unpublished data).

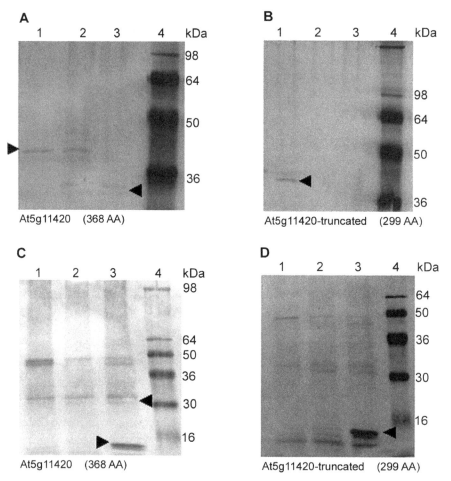

Fig. 7. 11420-protein *in vitro* interactors.

(ABCD) Affinity chromatography assays of 11420 interactors from DEAE-Sephacel flow-through protein fraction of *Arabidopsis thaliana* flowers (B,C) and leaves (D,E); Silver staining of 12% SDS-PAGE gels showing: (1) NaCl 100 mM, (2) NaCl 200 mM, (3) NaCl 500 mM elution fractions, and (4) molecular mass references.

(A) Flower interactors of the 11420 recombinant protein: Two protein bands are seen with molecular masses of 38 and 29 corresponding to FLR1 and VSP1 respectively (see arrows).

(B) Flower interactors of the 11420-truncated protein (Δ11420): In (1) only a 38kDa protein is detected, corresponding to FLOR1 (see arrow).

(C) Leaf interactors of the 11420 recombinant protein. In (1) and (2) two main protein bands are seen, corresponding to molecular masses of 37 and 32. In (3) Three proteins are detected with molecular masses of approximately 45, 32 and 14. The 32 kDa and 14 kDa bands were identified as the catalytic domain of a PME (see arrows).

(D) Leaf interactors of the 11420-truncated protein. The 14 kDa band shown in (3) (arrow) was identified as the same PME as in (C).

A

```
  1 MKLFVHLSIF FSILFITLPS SYSCTENDKN ALLQIKKALG NPPLLSSWNP RTDCCTGWTG
 61 VECTNRRVTG LSVTSGEVSG QISYQIGDLV DLRTLDFSYL PHLTGNIPRT ITKLKNLNTL
121 YLKHTSLSGP IPDYISELKS LTFLDLSFNQ FTGPIPGSLS QMPKLEAIQI NDNKLTGSIP
181 NSFGSFVGNV PNLYLSNNKL SGKIPESLSK YDFNAVDLSG NGFEGDAFMF FGRNKTTVRV
241 DLSRNMFNFD LVKVKFARSI VSLDLSQNHI YGKIPPALTK LHLEHFNVSD NHLCGKIPSG
```

B

```
  1 MKILSLSLLL LLAATVSHVQ SSASVPGLIE LLESNTIFGN EAELLEKEGL SINYPNCRSW
 61 HLGVETSNII NFDTVPANCK AYVEDYLITS KQYQYDSKTV NKEAYFYAKG LALKNDTVNV
121 WIFDLDDTLL SSIPYYAKYG YGTENTAPGA YWSWLESGES TPGLPETLHL YENLLELGIE
181 PIIISDRWKK LSEVTVENLK AVGVTKWKHL ILKPNGSKLT QVVYKSKVRN SLVKKGYNIV
241 GNIGDQWADL VEDTPGRVFK LPNPLYYVPS
```

C

```
  1 MAPSMKEIFS KDNFKKNKKL VLLSAAVALL FVAAVAGISA GASKANEKRT LSPSSHAVLR
 61 SSCSSTRYPE LCISAVVTAG GVELTSQKDV IEASVNLTIT AVEHNYFTVK KLIKRKGLT
121 PREKTALHDC LETIDETLDE LHETVEDLHL YPTKKTLREH AGDLKTLISS AITNQETCLD
181 GFSHDDADKQ VRKALLKGQI HVEHMCSNAL AMIKNMTDTD IANFEQKAKI TSNNRKLKEE
241 NQETTVAVDI AGAGELDSEG WPTWLSAGDR RLLQGSGVKA DATVAADGSG TFKTVAAVA
301 AAPENSNKRY VIHIKAGVYR ENVEVAKKK NIMFMGDGRT RTIITGSRNV VDGSTTFHSA
361 TVAAVGERFL ARDITFQNTA GPSKHQAVAL RVGSDFSAFY NCDMLAYQDT LYVHSNRQFF
421 VKCLIAGTVD FIFGNAAVVL QDCDIHARRP NSGQKNMVTA QGRTDPNQNT GIVIQKCRIG
481 ATSDLQSVKG SFPTYLGRPW KEYSQTVIMQ SAISDVIRPE GWSEWTGTFA LNTLTYREYS
541 NTGAGAGTAN RVKWRGFKVI TAAAEAQKYT AGQFIGGGGW LSSTGFPFSL GL
```

Fig. 8. Sequences of protein bands identified by LC-MS/MS from the pull down essays using DUF642 proteins. Bands were excised from the gels and sent to the Proteomics Platform of the Eastern Genomics Center, Quebec, Canada for their identification. One protein with high hit was identified for each band sent. Peptides identified are shaded. (A) FLR1 (Q9LH52, *At3g12145*) amino acid sequence showing all the peptides identified in different protein fractions. (B) VSP1 (Q93VJ6, *At5g24780*) amino acid sequence showing all the peptides identified in different protein fractions. (C) PME (Q9LUL7 *At3g14310*) amino acid sequence showing all the peptides identified in different protein fractions. Underlines show signal peptide (_ _ _), inhibitory domain (___) and catalytic region (___). Note that all the peptides identified for this protein match the catalytic domain.

Subcellular localisation is also an important criterion for putative *in vivo* protein interactions. Three bands of 37, 32 and 14 kDa were identified as fragments of the catalytic domain sequence from AtPME3 in leaf protein extracts. This electrophoretic pattern has been previously described in a purified citrus PME fraction. The enzymatic activity of the citrus PME fraction was not affected (Savary et al., 2002). However, this modification could be related to the subcellular localisation of AtPME3. The carboxylic 14 kDa fragment, which interacts with the *At5g11420*-encoded protein, was previously detected in the apoplastic fluid of rosette leaves (Boudart el al., 2005), whereas the complete AtPME3 catalytic domain that specifically interacts with 32460 protein was identified in the cell wall proteomes of different plant tissues (Feiz et al., 2006). FLOR1 was also detected in cell wall proteomes from different tissues.

The *in vitro* interactions of AtPME3 with the tested DUF642 proteins appear to be specific because no other PME was isolated with the affinity column. In particular, AtPME2 (*At1g53830*) shares a 90% sequence similarity to AtPME3, which is also present in the leaves. This result and the high similarity of the primary and secondary structures of both DUF642 proteins suggest that DUF642 proteins can interact with the same protein but with different isoforms that result from posttranslational modifications (Figures 6 and 7). A specific protein interaction of AtPME3 has been previously described. The cellulose-binding protein (CBP) secreted by the nematode *Heterodera schachtii* and that is involved in the infection process specifically interacts with AtPME3, and no interaction was detected with AtPME2 (Hewezi et al., 2008).

The interaction of PMEs with proteins is highly involved in cell wall remodelling. The interaction of PME with proteins that inhibit its activity contributes to the modulation of the methylesterified state of the pectin in the cell wall during different developmental processes (Pelloux et al., 2007). An important role of pectin modifications in the regulation of cell wall mechanics in the apical meristem tissue has been suggested (Peaucelle et al., 2011). In root tips, highly esterified pectins were found in the proliferating zone, and non-esterified pectins were abundant in the cell walls of differentiating cells (Barany et al., 2010). During pollen germination, the pollen tube wall presents highly methylesterified pectins in the tip region and weakly methylesterified pectins along the tube (Dardelle et al., 2010). It has been suggested that a local relaxation of the transmitting tract cell wall resulting from changes in the methylesterification of pectins could facilitate the growth of the pollen tubes in the extracellular matrix of this female tissue (Lehner et al., 2010).

4. Conclusions

The DUF642 domain contains a carbohydrate-binding module (CBM) that could be involved in cell wall polysaccharides. The presence of these modules has been described in enzymes from bacteria that hydrolyse hemicelluloses and pectins to degrade the plant cell wall (Kellet et al., 1990; Mc Kie et al., 2001). The function of these modules appears to be related to a precise targeting to polymers in specific regions of plant cell walls during developmental processes. Plant cell wall proteins can act as bridging proteins that target specific cell wall regions and crosslink different networks (Hervé et al., 2010). Additionally, 32460 and 11420 proteins interact *in vitro* with a PME and a LRR protein that are closely related to PGIPs. These two DUF642 proteins could be scaffold proteins that promote the complexation of PME and LRR proteins to prevent the targeting of non-esterified pectins by pectin-degrading enzymes such as polygalacturonases.

Our results suggest that FLOR1 and AtPME3 interact with the 11420 and 32460 DUF642 proteins, but the precise biochemical and biological functions remain to be determined.

5. Acknowledgments

This work was supported by PAPIIT grant IN220980 (Universidad Nacional Autónoma de México). EZS is a PhD student (Posgrado en Ciencias Biomédicas, UNAM) and received a CONACyT fellowship.

6. References

Abrahams, S., Hayes, C.M., & Watson, J. M. (1995). Expression pattern of three genes in the stem of Lucerne (*Medicago sativa*). *Plant Molecular Biology*, Vol. 27, No. 3, (February), pp 513-528, ISSN 0167-4412.

Acevedo, G.F., Gamboa, A., Páez-Valencia, J., Izaguirre-Sierra, M., & Alvarez-Buylla, R.E. (2004). FLOR1, a putative interaction partner of the floral homeotic protein AGAMOUS, is a plant-specific intracellular LRR. *Plant Science*, Vol.167, No. 2, (August), pp. 225-231, ISSN 0168-9452.

Albert, V.A., Soltis, D. E., Carlson, J.E., Farmerie, W.G., Kerr Wall, P., Ilut, D.C., Solow, T.M., Mueller, L.A., Landherr, L.L., Hu, Y.I., Buzgo, M., Kim, S., Yoo, M.J., Frohlich, M.W., Perl-Treves, R., Schlarbaum, S,E., Bliss, B.J., Zhang, X., Tanksley, S.D., Oppenheimer, D.G., Soltis, P.S., Ma, H., de Pamphilis, C.W., & Leebans-Mack, J.H. (2005). Floral gene resources from basal angiosperms for comparative genomics research. *BMC Plant Biology*, Vol. 5, No 5, (March), pp, ISSN 1471-2229.

Barcalá, M., García, A., Cabrera, J., Casson, S., Lindsey, K., Favery, B., García-Casado, G., Solano, R., Fenoll, C., & Escobar, C. (2010). Early transcriptomic events in microdissected *Arabidopsis* nematode-induced giant cells. *Plant Journal*, Vol. 61, No. 4, (February), pp. 698-712, ISSN 0960-7412.

Barany, I., Fadón, B., Risueño, M.C., & Testillano, P.S. (2010). Cell wall components and pectin esterification levels as markers of proliferation and differentiation events during pollen development and embryogenesis in *Capsicum annuum* L. *Journal of Experimental Botany*, Vol. 61, No. 4, (February), pp. 1159-1175, ISSN 0022-0957.

Baumberger, N., Doesseger, B., Guyot, R., Diet, A., Parsons, R.L., Clark, M.A., Simmons, M.P., Bedinger, P., Goff, S.A., Ringli, C., & Keller, B. (2003). Whole-genome comparison of leucine-rich repeat extensins in Arabidopsis and rice. A conserved family of cell wall proteins from vegetative and a reproductive clade. *Plant Physiology*, Vol. 131, No. 3, (March), pp. 1313-1326, ISSN 0032-0889.

Bayer, E.M., Bottrill, A.R., Walshaw, J., Vigouroux, M., Naldrett, M.J., Thomas, C.L., & Maule, A. (2006). Arabidopsis cell wall proteome defined using multidimensional protein identification technology. *Proteomics*, Vol. 6, No. 1, (January), pp. 301-311, ISSN 1615-9853.

Becnel, J., Natarajan, M., Kipp, A., & Braam, J. (2006). Developmental expression patterns of Arabidopsis XTH reported by transgenes and Genevestigator. *Plant Molecular Biology*, Vol. 61, No. 3, (June), pp. 451-467, ISSN 0167-4412.

Bischoff, V., Nita, S., Neumetzler, L., Schindelasch, D., Unrbain ,A., Eshed, R., Persson, S., Delmer, D., & Scheible, W.R. (2010a). TRICHOME BIREFRINGENCE and its homolog *At5g01360* encode plant-specific DUF231 proteins required for cellulose biosynthesis in Arabidopsis. *Plant Physiology*, Vol. 153, No. 2, (June), pp. 590-602, ISSN 0032-0889.

Bischoff, V., Selbig, J., & Scheible, W.R. (2010b). Involvement of TBL/DUF231 proteins into cell wall biology. *Plant Signaling & Behavior*, Vol. 5, No. 8, (August), pp. 1057-1059, ISSN 1559-2316.

Borderies, G., Jamet, E., Lafitte, C., Rossignol, M., Jauneau, A., Boudart, G., Monsarrat, B., Esquerré-Tugayé, M.T., Boudet, A., & Pont-Lezica, R. (2003). Proteomics of loosely bound cell wall proteins of *Arabidopsis thaliana* cell suspension cultures: a critical analysis. *Electrophoresis*, Vol. 24, No. 19-20, (October), pp. 3421-3432, ISSN 0173-0835.

Borner, G.H., Lilley, K.S., Stevens, T.J., & Dupree, P. (2003). Identification of glycosilphosphatidylinositol anchored proteins in Arabidopsis. A proteomic and genomic analysis. *Plant Physiology*, Vol. 132, No. 2, (June), pp. 568-577, ISSN 0032-0889.

Boudart, G., Jamet, E., Rossignol, M., Lafitte, C., Borderies, G., Jauneau, A., Esquerré-Tugayé, M.T., & Pont-Lezica, R. (2005). Cell wall proteins in apoplastic fluids of *Arabidopsis thaliana* rosettes: Identification by mass spectometrometry and bioinformatics. *Proteomics*, Vol. 5, No.1 , (January), pp. 212-221, ISSN 1615-9853.

Cassab, G.I. (1998). Plant cell wall proteins. *Annual Reviews of Plant Physiology and Plant Molecular Biology*, Vol. 49, pp. 281-309, ISSN 1040-2519.

Charmont, S., Jamet, E., Pont-Lezica, R., & Canut, H. (2005). Proteomic analysis of secreted proteins from *Arabidopsis thaliana* seedlings: improved recovery following removal of phenolic compounds. *Phytochemistry*, Vol. 66, No. 4, (February), pp. 453-461, ISSN 0031-9422.

Chibani, K., Ali-Rachedi, S., Job, C., Job, D., Jullien, M., & Grappin, P. (2006). Proteomic analysis of seed dormancy in Arabidopsis. *Plant Physiology*, Vol.142, No. 4, (December), pp. 1493-1510, ISSN 0032-0889.

Chivasa, S., Ndimba, B.K., Simon, W.J., Robertson, D., Yu, X.L., Knox, J.P., Bolwell, P., & Slabas, A.R. (2002). Proteomic analysis of *Arabidopsis thaliana* cell wall. *Electrophoresis*, Vol. 23, No. 11, (Junio), pp. 1754-1765, ISSN 0173-0835.

Clark, S.E., Williams, R.W., & Meyerowitz, E.M. (1997). The *CLAVATA1* gene encodes a putative receptor kinase that controls shoot and meristem size in Arabidopsis. *Cell*, Vol. 89, No. 4, (May), pp. 575-585, ISSN 0092-8674.

Dardelle, F., Lehner, A., Ramdani, Y., Basrdor, M., Lerouge, P., Driouich, A., & Mollet, J.C. (2010). Biochemical and immunocytological characterizations of Arabidopsis pollen tube cell wall. *Plant Physiology*, Vol. 153, No. 4, (August), pp. 1563-1576, ISSN 0032-0889.

Depuydt, S., Trenkamp, S., Fernie, A.R., Elftieh, S., Renou, J.P., Vuylsteke, M., Holsters, M., & Vereecke, D. (2009) An integrated genomics approach to define niche establishment by *Rhodococcus fascians*. *Plant Physiology*, Vol. 149, No. 3, (March), pp. 1366-1386, ISSN 0032-0889.

Dhonukshe, P., Baluska, F., Schlicht, M., Hlavacka, A., Samaj, J., Friml, J., & Gadella, T.W.J (2006). Endocytosis of cell surface material mediates cell plate formation during plant cytokinesis. *Developmental Cell*, Vol. 10, No. 1 (January), pp. 137-150, ISSN 1534-5807.

Di, C., Zhang, M., Xu, S., Cheng, T., & An, L. (2006). Role of poly-galacturonase inhibiting protein in plant defense. *Critical Reviews of Microbiology*, Vol. 32, No. 2, (January), pp 91-100, ISSN 1040-841X.

Dunkley, T.P., Hester, S., Shadforth, I.P., Runions, J., Weimar, T., Hamton, S.L., Griffin, J.L., Bessant, C., Brandizzi, F., Hawes, C., Watson, R.B., Dupree, P., & Lilley, K.S. (2006). Mapping the Arabidopsis organelle proteome. *Proceedings of the National Academy of Science*,Vol. 103, No. 11, (April), pp. 1128-1134, ISSN 1091-6490.

Ehlting, J., Chowrira, S.G., Matthews, N., Eschliman, D.S., Arimura, G., & Bohlmann, J. (2008). Comparative transcriptome analysis of *Arabidopsis thaliana* infested by diamond back moth (*Plutella xylostella*) larvae reveals signatures of stress response, secondary metabolism, and signalling. *BMC Genetics*, Vol. 9, No. 9 (April), pp. 154-160, ISSN 1471-2156.

Feiz, L., Irshad, M., Pont-Lezica, R.F., Canut, H., & Jamet, E. (2006). Evaluation of cell wall preparations for proteomics: A new procedure for purifying cell walls from Arabidopsis hypocotyls. *Plant Methods*, Vol. 2, No. 10, (May), pp. 13, ISSN 1746-4811.

Feraru, E., Feraru, M.I., Kleine-Vehm, J., Martiniere, A., Mouille, J., Vanneste, S., Vernhettes, S., Runions, J., & Friml, J. (2011). PIN polarity maintenance by the cell wall in Arabidopsis. *Current Biology*, Vol. 21, No. 4, (February), pp. 338-343, ISSN 0960-9822.

Gaboriaud, C., Bissery, V., Benchetrit, T., & Mornon, J.P. (1987) Hydrophobic cluster analysis: an efficient new way to compare and analyse amino acid sequences. *FEBS Letters*, Vol. 224, No. 1, (November), pp. 149-155, ISSN 0014-5793.

Gamboa, A., Páez-Valencia, J., Acevedo, G.F., Vázquez-Moreno, L., & Alvarez-Buylla, R.E. (2001). Floral transcription factor AGAMOUS interacts in vitro with a leucine-rich repeat and an acid phosphatase complex. *Biochemical and Biophysics Research Communications*, Vol. 288, No. 4, (November), pp. 1018-1026, ISSN 0006-291X.

Goda, H., Sawa, S., Asami, T., Fujioka, S., Shimada, Y., & Yoshida, S. (2004). Comprehensive comparison of auxin-regulated and brassinosteroid regulated genes in Arabidopsis. *Plant Physiology*, Vol. 134, No. 4, (April), pp. 1555-1573, ISSN 0032-0889.

González-Carranza, Z.H., Elliot, K.A., & Roberts, J.A. (2007). Expression of polygalacturonases and evidence to support its role Turing cell separation processes in *Arabidopsis thaliana*. *Journal of Experimental Botany*, Vol 58, No. 13, (October), pp. 3719-3730, ISSN 0032-0889.

Grunewald, W., van Noorden, G., Van Isterdaal, G., Beeckman, T., Gheysen, G., & Mathesius, U. (2009). A manipulation of auxin transport in plant roots during Rhizobium simbiosis and nematode parasitism. *Plant Cell* Vol. 21, No. 9, (September), pp. 2553-2562, ISSN 1040-4651.

Guenin, S., Mareck, A., Rayon, C., Lamour, R., Assoumou, Ndong, Y., Domon, J.M., Senechal, F., Fourner, F., Jamet, E., Canut, H., Percoco, G., Moville, G., Rolland, A., Rusterucci, C., Guerineau, F., Van Wuytswinkel, O., Gillet, F., Driovich, A., Lerouge, P., Gutierrez, L, & Pelloux, J. (2011). Identification of pectin methylesterase 3 as a basic pectin methylesterase isoform involved in adventitious rooting in *Arabidopsis thaliana*. *New Phytologist*, Vol. 192, No. 1, (October), pp. 114-126, ISSN 0028-646X.

Harpster, M.H., Brummell, D.A., & Dunsmuir, P. (2002). Suppression of a ripening-related endo-1,4-β-glucanase in transgenic pepper fruit does not prevent depolymerisation of cell wall polysaccharides during ripening. *Plant Molecular Biology*, Vol. 50, No. 3, (October), pp. 345-355, ISSN 0167-4412.

Havlis, J., Thomas, H., Sebela, M., & Shevchenko, A. (2003). Fast-response proteomics by accelerated in-gel digestion of proteins. *Analytical Chemistry*, Vol. 75, No. 6, (March), pp. 1300-1306, ISSN 0003-2700.

Held, M.A., Be, E., Zemelis, S., Withers, S., Wilkerson, C., & Brandizzi, F. (2011). CRG3: A golgi-localized protein influencing homogalacturonan methylesterification. *Molecular Plant*, Vol. 4, No. 5, (September), pp. 832-844, ISSN 1674-2052.

Hervé, C., Rogowski, A., Blake, A.W., Marcus, S.E., Gilbert, H.J., & Knox, J.P. (2010). Carbohydrate-binding modules promote the enzymatic deconstruction of intact plant cell walls by targeting and proximity effects. *Proceedings of the National Academy of Science*, Vol. 107, No. 34, (August), pp 15293-15298, ISSN 1091-6490.

Hewezi, T., Howe, P., Maier, T,R., Hussey, R.S., Mitchum, M.G., Davis, E.L., & Baum, T.J. (2009). Cellulose binding protein from the parsitic nematode *Heterodera schachtii* interacts with Arabidopsis pectin methyl-esterase: cooperative cell-wall modification during parasitism. *Plant Cell*, Vol. 20, No. 11, (November), pp. 3080-3093. ISSN 1040-4651.

Hoffman, R.M., and Turner, J.G. (1984), Occurrence and specificity of an endopolygalacturonase inhibitor in *Pisum sativum*. *Physiol. Plant Pathol.* Vol. 24, No. 1, (January), pp. 49-59, ISSN 0048-4059.

Hu, J., Barlet, X., Deslandes, L., Hirsch, J., Feng, X., Somssich, I., & Marco, Y. (2008). Transcriptional responses of *Arabidopsis thaliana* during wilt disease caused by soilborne phytopathogenic bacterium *Ralstonia solanacearum*. *PLoS ONE*, Vol. 3, No. 7, (July), e2589, ISSN 1932-6203.

Irshad, M., Canut, H., Borderies, G., Pont-Lezica, R., & Jamet, E. (2008). A new picture of cell wall protein dynamics in elongating cells of *Arabidopsis thaliana*; confirmed actors and new comers. *BMC Plant Biology*, Vol. 8, (September), pp. 94-103.

Jamet, E., Canut, H., Boudart, G., & Pont-Lezica, R.F. (2006). Cell wall proteins: a new insight through proteomics. *Trends in Plant Sciences*, Vol. 11, No. 1, (January), pp. 33-39. ISSN 1360-1385.

Juge, N (2006). Plant proteins inhibitors of cell wall degrading enzymes. *Trends in Plant Sciences*. Vol. 11, No. 7, (July), pp. 359-367. ISSN 1360-1385.

Kajava, A.V. (1998). Structural diversity of Leucine-rich repeat proteins. *Journal of Molecular Biology*, Vol. 277, No. 3, (April), pp. 519-527, ISSN 0022-2836.

Keller, A., Nesvizhskii, A.I., Kolker, E., & Aebersold, R. (2002). Empirical statistical model to estimate the accuracy of peptide identifications made by MS/MS and database search. *Analytical Chemistry*, Vol. 74, No. 20, (October), pp. 5383-5392, ISSN 0003-2700.

Kellet, L.E., Poole, D.M., Ferreira, L.M., Durrant, A.J., Hazlewood, G.P., & Gilbert, H.J. (1990). Xylanase B and arabinofuranosidase from *Pseudomonas fluorescens* subsp. cellulosa contain identical cellulose-binding domains and are encoded by adjacent genes. *Biochemical Journal*, Vol. 272, No. 2, (December), pp. 369-376. ISSN 0244-6021.

Kreps, J.A., Wu, Y., Chang, H-S., Zhu, T., Wang, X., & Harper, J.F. (2002). Transcriptome changes for Arabidopsis in response to salt, osmotic, and cold stress. *Plant Physiology*, Vol. 130, No. 4, (December), pp. 2129-2141, ISSN 0032-0889.

Lannoo, N., Vanderborre, G., Miersch, O., Smagghe, G., Wasternack, C., Peumans, W.J., & Van Damme, E.J. (2007). The jasmonate-induced exporession of the *Nicotiana tabacum* in leaf lectins. *Plant and Cell Physiology*, Vol. 48, No. 8, (August), pp. 1207-1218, ISSN 0032-0781.

Lee, C.W., Efetova, M., Engelmann, J.C., Kramell, R., Wastermack, C., Ludwig-Müller, J., Hedrich, R., & Deeken, R. (2009). *Agrobacterium tumefaciens* promotes tumor induction by modulating pathogen defense in *Arabidopsis thaliana*. *Plant Cell*, Vol. 21, No. 9, (September), pp. 2948-2962, ISSN 1040-4651.

Lehner, A., Dardelle, F., Soret-Morvan, O., Lerouge, P., Driouich, A., & Mollet, J.C. (2010). Pectins in the cell wall of *Arabidopsis thaliana* pollen tube and pistil. *Plant Signaling & Behavior*, Vol. 5, No. 10, (October), pp. 1282-1285.

Mc Kie, V.A., Vincken, J.O., Voragen, A.G., van der Broek, L.A., Stimson, E., & Gilbert H.J. (2001). A new family of rhamnogalacturonan lyases contains an enzyme that binds to cellulose. *Biochemical Journal*, Vol. 355, No. 1, (April), pp. 167-177. ISSN 0244-6021.

Membré, N., Bernier, F., Staiger, D., & Berna, A. (2000). *Arabidopsis thaliana* germin-like proteins: common and specific features point to a variety of functions. *Planta*, Vol. 211, No. 3, (August), pp.345-354, ISSN 0032-0935.

Micheli, F. (2001). Pectin methylesterases : cell wall enzymes with important roles in plant physiology. *Trends in Plant Science*, Vol. 6, No. 9, (September), pp. 414-418. ISSN 1360-1385.

Minic, Z., Jamet, E., Négroni, L., der Garabedian, A., Zivy, M., & Jouanin, L. (2007). A sub-proteome of *Arabidopsis thaliana* mature stems trapped on Concanavalin A is enriched in cell wall glycoside hydrolases. *Journal of Experimental Botany*, Vol. 58, No. 10, (May), pp. 2503-2512, ISSN 0022-0957.

Passardi, F., Penel, C., & Dunand, C. (2004). Performing the paradoxical: how plant peroxidises modify the cell wall. *Trends in Plant Science*, Vol. 9, No. 11, (November), pp. 534-540, ISSN 1360-1385.

Passardi, F., Tognolli, M., De Meyer, M., Penel, C., Dunand C., (2006). Two cell wall associated peroxidases from Arabidopsis influence root elongation. *Planta*, Vol. 223, No. 5, (April), pp. 965-974, ISSN 0032-0935.

Peaucelle, A., Braybrook, S.A., Le Guillou, L., Bron, E., Kuhlemeier, C., & Höfte, H. (2011). Pectin induced changes in cell wall mechanics underlie organ initiation in Arabidopsis. *Current Biology*, Vol. 21, No. 20, (October), 1720-1726, ISSN 0960-9822.

Pelloux, J., Rustérucci, C., & Mellerowicz, E.J. (2007). New insights into pectin methylesterase structure and function. *Trends in Plant Science*, Vol. 12, No. 6, (June), pp. 267-277, ISSN 1360-1385.

Reinhardt, D., Wittwer, F., Mandel, T., & Kuhlemeier, C. (1998). Localized upregulation of a new expansin gene predict the site of leaf formation in the tomato meristem. *Plant Cell*, Vol. 10, No. 9, (September), pp. 1427-143,. ISSN 1040-4651.

Röckel, N., Wolf, S., Kost, B., Rausch, T., & Greiner, S. (2008). Elaborate spatial patterning of cell wall PME and PMEI at the pollen tube tip involves PMEI endocytosis and reflects the distribution of esterified and de-esterified pectin. *Plant Journal*, Vol. 53, No. 1, (January), pp. 133-143, ISSN 0960-7412.

Ruppert, M., Woll, J., Giritch, A., Genady, E., Ma, X., & Stockigt, J. (2005). Functional expression of an ajmaline pathway-specific esterase from Rauvolfia in a plant-virus expression system. *Planta*, Vol. 222, No. 5, (November), pp. 888-898, ISSN 0032-0935.

Sampedro, J., & Cosgrave, D.J. (2005). The expansin superfamily. *Genome Biology*, Vol. 6, No. 12, (November), pp.242-250, ISSN 1474-7596.

Savary, B.J., Hotchkiss, A.T., & Cameron, R.G. (2002). Characterization of a salt independent pectin methylesterase purified from Valencia orange peel. *Journal of Agriculture and Food Chemistry*, Vol. 50, No. 12, (June), pp. 3553-3558, ISSN 0021-8561.

Shevchenko, M., Wilm, O., Vorm, O., & Mann, M. (1996). Mass spectrometric sequencing of proteins from silver-stained polyacrylamide gels. *Analytical Chemistry*, Vol. 68, No. 5 (March), pp. 850-858, ISSN 0003-2700.

Sterling, J.D., Quigley, H.F., Oerallana, A., & Mohnen, D. (2001). The catalytic site of the pectin biosynthetic enzyme alpha-1,4-galacturonosyltransferases is located in the lumen of the Golgi. *Plant Physiology*, Vol. 127, No. 1, (September), pp. 360-371, ISSN 0032-0889.

Thomas, C.L., Bayer, E.M., Ritzenthaler, C., Fernandez-Calvino, L., & Maule, A.J. (2008). Specific targeting of a plasmodesmal protein affecting cell-to-cell communication. *PLoS Biology*, Vol. 6, No.1, (January), e7, ISSN 1932-6203.

Tung, C.W., Dwyer, K.G., Nasrallah, M.E., & Nasrallah, J.B. (2005). Genome-wide identification of genes expressed in Arabidopsis pistils specifically along the path of pollen tube growth. *Plant Physiology*, Vol. 138, No. 2, (June), pp. 977–998, ISSN 0032-0889.

Van Damme, E.J., Lanoo, N., Fouquaert, E., & Peumans, W.J. (2004). The identification of inducible cytoplasmic/nuclear carbohydrate-binding proteins urges to develop novel concept about the role of plant lectins. *Glycoconjugate Journal*, Vol. 20, No. 7-8, pp. 449-460. ISSN 0282-0080.

van der Hoorn, R.A.L. (2008). Plant proteases: from phenotypes to molecular mechanisms. *Annual Review of Plant Biology*, Vol. 59, No. 191, (June), pp. 191-223, ISSN 1543-5008.

Wellmer, F., Riechemann, J.L., Alves-Ferreira, M., & Meyerowitz, E.M. (2004). Genome wide analysis of spatial gene expression in Arabidopsis flowers. *Plant Cell*, Vol. 16, No. 5 (May), pp. 1314- 1326, ISSN 1040-4651.

Wolf, S., Mouille, G., & Pelloux, J. (2009). Homogalacturonan methyl-esterification and plant development. *Molecular Plant* Vol. 5, No. 2 (September), pp. 851-860, ISSN 1674-2052.

Wraczeck, M., Brosché, M., Salojärvi, J., Kangasjärvi, S., Idänheimo, N., Mersmann, S., Robatzek, S., Karpinski, S., Karspinska, B., & Kangasjärvi, J. (2010). Transcriptional regulation of CRK/DUF26 group of receptor-like protein kinases by ozone and plant hormones in Arabidopsis. *BMC Plant Biology*, Vol. 10, No. 95, (May), pp 95-101, ISSN 1471-2229-

Wu, H.C., Hsu, S.F., Luo, D.L., Chen, S.J., Huang, W.D., Lur, H.S., & Jinn, T.L. (2010). Recovery of heat shock-triggered released apoplastic Ca2+ accompanied by pectin methylesterase activity is required for thermotolerance in soybean seedlings. *Journal of Experimental Botany*, Vol. 61, No. 10, (June), pp. 2843-2852, ISSN 0022-0957.

6

AApeptides as a New Class of Peptidomimetics to Regulate Protein-Protein Interactions

Youhong Niu[1,*], Yaogang Hu[1,*], Rongsheng E. Wang[1,*],
Xiaolong Li[2,*], Haifan Wu[1], Jiandong Chen[2,**] and Jianfeng Cai[1,**]

[1]Department of Chemistry, University of South Florida, Tampa, FL
[2]Department of Molecular Oncology, H. Lee Moffitt Cancer Center & Research Institute,
Tampa, FL
USA

1. Introduction

In human physiology, proteins at all times are synthesized, processed, degraded and post-translationally modified to varying degrees and at different rates, for their participations in a wide variety of activities to maintain normal functions of the body (Murray et al. 2007). In the successful proceedings of all the biological events, signals are consistently received and sent via physical contacts between proteins (Murray et al. 2007). The communication between proteins or the alternatively called non-covalent protein-protein interactions are thereby considered as important as proteins' own functions (Murray et al. 2007). Disruptions of these signaling pathways by either mutational changes or deregulation of one of the protein partners would result in a series of diseases (Murray et al. 2007). On the other hand, therapeutic approaches based on chemical agents could potentially inhibit protein-protein interactions, thereby restoring the balance of signaling pathways, and leading to the cure of diseases (Murray et al. 2007). However, it is quite challenging to develop chemical agents which can target protein-protein interactions (Arkin et al. 2004; Whitty et al. 2006). Unlike the traditional medicinal chemistry approach, in which small molecule inhibitors are developed to target the hydrophobic pocket of enzymes / kinases, chemical agents are now required to bind to large surfaces of proteins that are usually amphiphilic and flexible (Murray et al. 2007). Yet a number of successful stories have been reported (Murray et al. 2007). Taking the p53/MDM2 system as an example, the p53/MDM2 has been a model system for the inhibition of protein-protein interactions, and has been reported to be the targets of a wide variety of inhibitors (Oren 1999; Balint et al. 2001; McLure et al. 2004; Brooks et al. 2006).

The tumor suppressor protein p53 is a transcription factor that executes multiple anticancer functions. Through its binding to DNA, p53 can initiate the expression of several important proteins, which are responsible for DNA repair, induction of growth arrest to hold the cell

---

* These authors contributed to this work equally
** Corresponding author

cycle at the G1/S regulation point, as well as the initiation of apoptosis (Lowe et al. 1993; Pellegata et al. 1996; Liu et al. 2001). However, at the normal state, the p53 activity is down-regulated by the murine double minute 2 protein (MDM2) which binds to the α-helical transactivation domain near the N-terminus of p53 (Momand et al. 1992; Oliner et al. 1992). Cocrystal structure studies revealed that three hydrophobic side chains from Phe19, Trp23, and Leu26 of p53 make direct contacts with MDM2 and account for the primary interactions (Kussie et al. 1996). The binding of MDM2 not only inhibits p53 DNA-binding activity but also induces the proteosomal degradation of p53 (Haupt et al. 1997; Kubbutat et al. 1997) . In the event of stress, p53 protein is phosphorylated, which leads to a much reduced affinity between p53 and MDM2, and thereby reactivating p53 (Jimenez et al. 1999). Nonetheless, MDM2 is constantly over-expressed in tumor cells, which significantly blocks the activation of p53 pathway even during stress conditions, thereby leading to the uncontrolled tumor cell proliferation (Momand et al. 1998). The overproduction of MDM2 makes tumors less susceptible to programmed cell death and apoptosis as a result of chemotherapy and other cancer therapy. Hence, the disruption of MDM2/p53 interaction in tumor cells should stabilize p53, preventing it from degradations, and initiating a cascade of p53 pathways to eventually sensitize the tumor cells to death (Murray et al. 2007). To date, targeting the MDM2/p53 interaction has become an emerging therapeutic approach in anticancer treatment. Numerous efforts have been taken for the development of inhibitors such as natural products (De Vincenzo et al. 1995; Stoll et al. 2001; Duncan et al. 2003; Tsukamoto et al. 2006), small molecules (Zhao et al. 2002; Galatin et al. 2004; Vassilev et al. 2004; Ding et al. 2005; Grasberger et al. 2005; Hardcastle et al. 2006), and oligomers (Alluri et al. 2003; Hara et al. 2006; Robinson 2008; Hayashi et al. 2009; Michel et al. 2009; Bautista et al. 2010). Compared to small molecule approach, oligomers are easily programmable and are readily synthesized by solid phase synthesis. It is also believed that the larger size of oligomers relative to small molecules may bring them additional advantages to contact more protein surface area, which will lead to enhanced binding affinity (Murray et al. 2007).

However, oligomers made of natural peptides are subjected to biodegradations and are also immunogenic in vivo, which limit their practical applications, but on the other hand underscore the need for unnatural peptides (Patch et al. 2002). Peptidomimetics are a class of non-natural peptide mimics using the artificial backbones to mimic peptides' primary and secondary structures (Wu et al. 2008). Compared to traditional peptides, peptidomimetics have great proteolytic and metabolic stability and are believed to be less immunogenic, also with an enhanced bioavailability (Patch et al. 2002; Goodman et al. 2007; Wu et al. 2008). The development of peptidomimetics to disrupt MDM2/p53 has led to a diverse set of oligomers such as β-peptides (Seebach et al. 1996; Cheng et al. 2001; Kritzer et al. 2005), γ- and δ-peptides (Arndt et al. 2004; Trabocchi et al. 2005; Kumbhani et al. 2006), α/β-peptides (Horne et al. 2008; Horne et al. 2009), azapeptides (Graybill et al. 1992; Lee et al. 2002), α-aminoxy-peptides (Li et al. 2008), sugar-based peptides (Risseeuw et al. 2007; Tuwalska et al. 2008), peptoids (Simon et al. 1992), oligoureas (Boeijen et al. 2001; Violette et al. 2005), polyamides (Dervan 1986), and phenylene ethynylenes (Nelson et al. 1997), etc. Nonetheless, the development of peptidomimetics is far less straightforward, with the major limit lying in the availability of framework (Goodman et al. 2007). The search for peptidomimeitcs of a variety of backbones remains crucial in the research of peptide mimics, which would result in different classes of oligomers with diverse structures and functions (Goodman et al. 2007;

Horne et al. 2008; Gellman 2009). The development of new peptidomimetics would also facilitate the identification of novel therapeutic agents and help the understanding of protein folding and functions by using peptidomimetic probes, all of which are important to the progress in modern chemical biology research (Goodman et al. 2007; Horne et al. 2008; Gellman 2009).

2. Development of AApeptides

In the attempt to search for new peptide mimics for drug discovery and protein mimicry, we recently described a novel class of peptidomimetics termed "AApeptides", which is derived from N-acylated-N-aminoethyl amino acids that has been previously used as the building block for PNA (Winssinger et al. 2004; Dragulescu-Andrasi et al. 2006; Debaene et al. 2007). Compared to natural peptides, the repeating unit of the AApeptide is structurally similar to two adjacent residues of α-peptide, in which there are two side chains, one from the regular α-amino acid side chain, while the other one from a carboxylic acid residue appended to the tertiary amide nitrogen. Depending on the relative position of α-amino acid side chain, there are two types of AApeptides. The one with α-amino acid side chain at the α position is called α-AApeptide (Hu et al. 2011), while the other one with side chain at the γ position is called γ-AApeptide (Niu et al. 2011) (Figure 1).

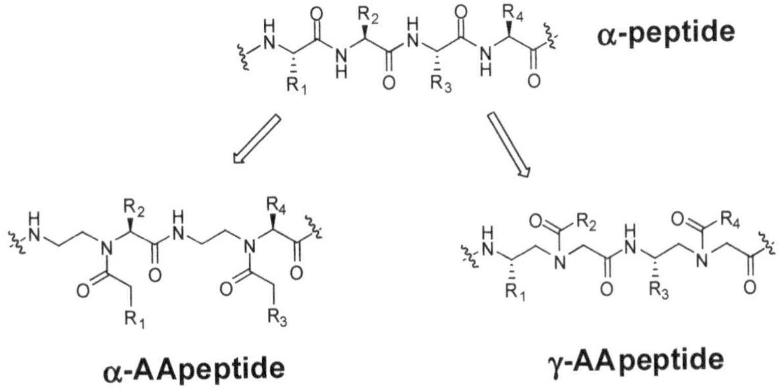

Fig. 1. Structures of a α-peptide and the corresponding AApeptides.

Both type of AApeptides project the same number of functional groups as conventional peptides with backbones of the same length. In addition, all the nitrogen atoms of AApeptides are involved in either secondary or tertiary amide bonds, in a way similar to natural α-peptide. Taken together, such AApeptides are designed to mimic the distance relationships and relative positions of amino acid side chains of natural peptides, so that they can reserve some functions of conventional peptides. It is also noteworthy that, even though AApeptides can mimic the structure as well as some activities of natural peptides, they are still different in backbone and should possess distinct hydrogen bonding properties and conformational flexibilities. The backbone of AApeptide is more flexible, with involved tertiary amide bonds potentially in cis/trans conformations, suggesting that the direct inter-conversion of sequences between AApeptides and natural peptides may not result in the same activity and functions.

3. Development of AApeptides for inhibition of p53/MDM2 interaction

For proof of concept, we demonstrated the facile synthesis and potential bioactivities of AApeptides by developing AApeptide based inhibitors of the p53/MDM2 model system, which has been a testing ground for freshly developed peptidomimetics of novel frameworks.

3.1 Design of AApeptide sequences

Previous reports indicated that synthetic agents displaying hydrophobic side chains of Phe19, Trp23, and Leu26, and in the orientation mimicking the array of these amino acids in p53 should compete with p53 in occupying the MDM2 cleft (Murray et al. 2007). Based on these findings, we designed four α-AApeptides and three γ-AApeptides to mimic the binding surface of p53 (Figure 2 and 3).

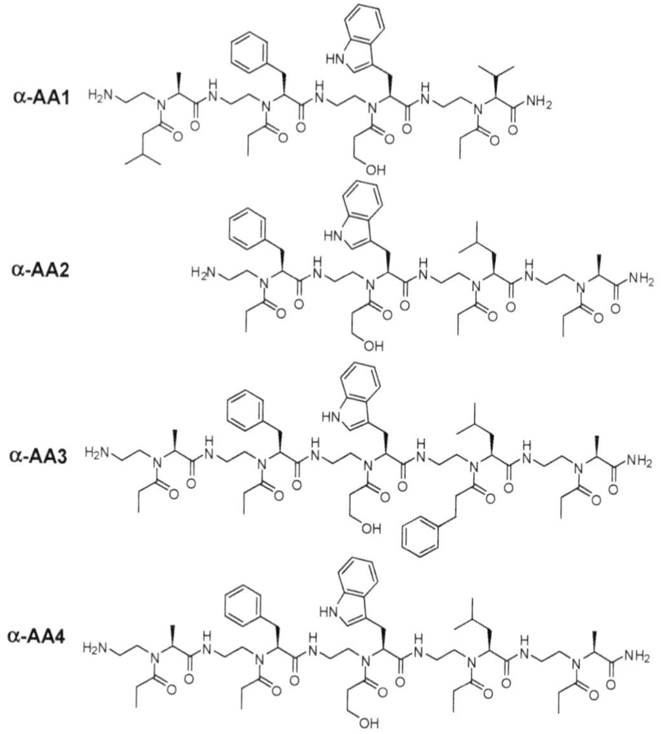

Fig. 2. α-AApeptide sequences designed for inhibition of p53 / MDM2 interaction. Figure is adapted from (Hu et al. 2011).

These AApeptides bear either some or all of the functional side chains of the three amino acids (Phe19, Trp23, and Leu26), which are designed to be the amino acid side chains at either α (for α-AApeptides) or γ positions (for γ-AApeptides). The other functional groups were randomly chosen, with most of them appended to the nitrogens of AApeptides through the formation of tertiary amide bonds.

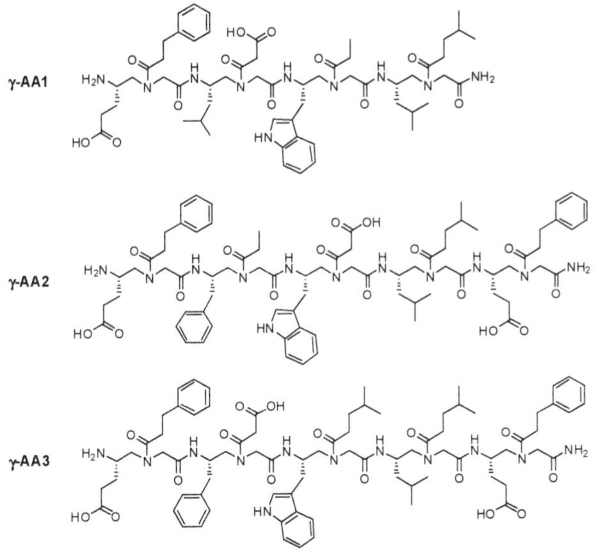

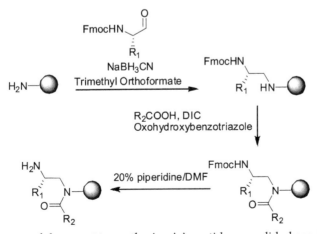

Fig. 3. γ-AApeptide sequences designed for p53/MDM2 disruption. Figure is adapted from (Niu et al. 2011).

3.2 Synthesis of AApeptides

In our initial attempt, we tried to synthesize AApeptides on solid phase resins through a direct sub-monomer strategy, in which the functional groups were introduced to the sequence step by step (Figure 4). Unfortunately, the presence of multiple secondary amines in the peptide backbone led to a constant over-alkylation during the reductive amination step. As a result, we only obtained a mixture of unidentified products after several coupling

Fig. 4. Initial unsuccessful attempt to synthesize AApeptides on solid phase. Figure is adapted from (Niu et al. 2011).

cycles, as observed on HPLC after the cleavage of products from solid phase. We then carried out an alternative "monomer building block" strategy, in which building block was first synthesized in solution phase, and then assembled following the same procedure of standard solid phase synthesis of conventional peptides. In this route, AApeptide building blocks are readily prepared using commercially available agents at low cost.

For the synthesis of α-AApeptide building block (Figure 5), the carboxylic acid of amino acid was first protected to form the amino acid benzyl ester (**A**) or the amino acid tert-butyl ester (**B**). The resulting amino acid esters were then reacted with Fmoc-amino ethyl aldehyde by reductive amination to form secondary amines **2**, which were subsequently acylated with functional groups R^1. The coupling products **3** were finally deprotected with hydrogenlysis to remove the benzyl protecting group, or with trifluoroacetic acid to remove the tert-butyl protecting group. For the synthesis of γ-AApeptide building block (Figure 6), glycine benzyl ester was reacted with Fmoc-amino aldehydes through reductive amination, and the resulting intermediates **2** was subsequently acylated with carboxylic acid ended functional groups to form the coupled intermediate **3**. After a hydrogenation step, the desired γ-AApeptide building blocks **4** were finally obtained. For both α and γ AApeptides, a diverse set of conjugation conditions for the preparation of intermediates **3** were investigated. It was found that the use of activation agents such as HBTU/HOBt, DIC/HOBt, or PyBOP can provide the desired products in poor yields, and only when intermediate **2** was conjugated with a few types of carboxylic acids. After many trials, the coupling with oxohydroxybezotriazole / DIC emerged as the most efficient and can catalyze the successful conjugations of most carboxylic acids. It is also noteworthy that the derivatization of AApeptides is virtually limitless, since there are countless carboxylic acids available for acylation of nitrogen atom in the backbone. This specific feature allows the rapid generation of AApeptide library, which literally should have much more diversity than those libraries based on regular peptides, thereby expanding the versatility of oligomer libraries for potential applications in high-throughput screening based drug discovery and chemical biology research. With the prepared building blocks in hand, the solid phase synthesis of AApeptides was carried out on resins in a simple and highly efficient way. The sequences were finally obtained over 80% yield in crude and were purified by HPLC to achieve purities over 95%. Their identities were further confirmed by MALDI-mass spectrometry.

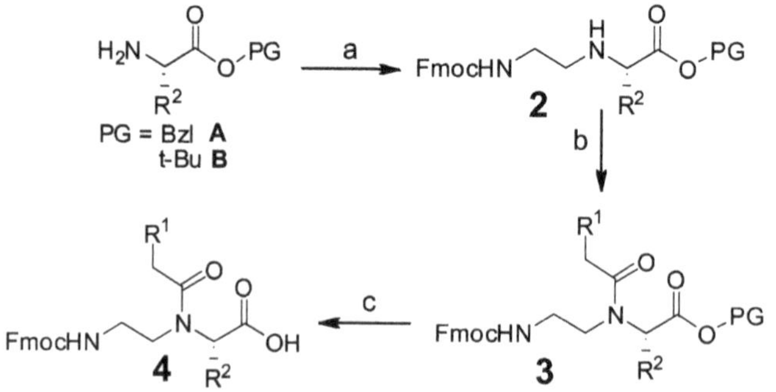

Fig. 5. Synthesis scheme of α-AApeptide building block. a) Fmoc-amino ethyl aldehyde, NaBH₃CN, overnight. b) R_1CH_2COOH, DhBtOH/DIC, overnight. c) Pd/C, H_2 for A; 50% TFA/CH_2Cl_2 for B. Figure is adapted from (Hu et al. 2011).

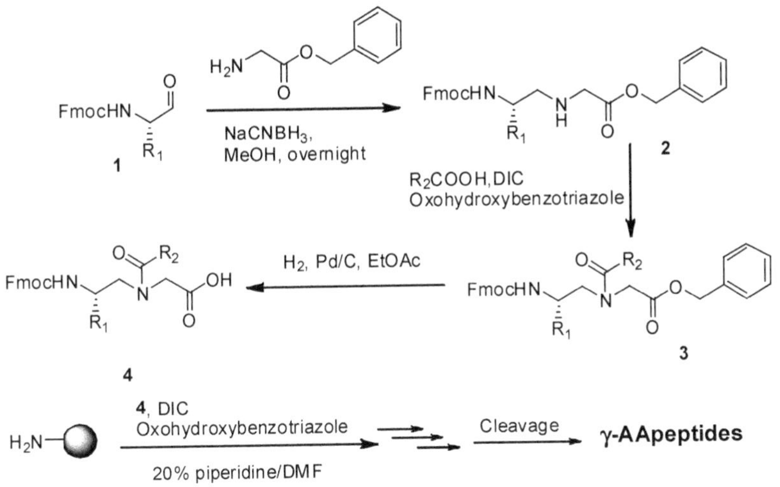

Fig. 6. Synthesis scheme of γ-AApeptide building block. Figure is adapted from (Niu et al. 2011).

3.3 ELISA assay of AApeptides for inhibition of p53/MDM2 interaction

These AApeptides were then tested by the ELISA assay for their inhibition of p53/MDM2 protein-protein interaction. Generally the ELISA plate was coated with p53, and then incubated with the mixture of MDM2 protein and AApeptide for one hour. The p53 bound MDM2 protein was then detected by MDM2 antibody and a secondary antibody conjugated with the horseradish peroxidase, which later on reacted with the TMB peroxidase substrate to manifest a yellowish color after acid quenching. The color intensity was monitored by absorbance at 450nm, the extent of which is directly proportional to the amount of bound MDM2 protein, and also reversely correlated to the inhibiting efficiency of AApeptide. The readings for each sample were then plotted against the concentration of α-AApeptide (Figure 7) or γ-AApeptide (Figure 8). Almost all AApeptides inhibit the p53-MDM2 binding when they are administrated at high concentrations.

Based on the plots of each peptide in Figure 7 and 8, the related IC_{50} values were calculated and summarized in Table 1. The published IC_{50} value of the wide type p53-drived peptide (Garcia-Echeverria et al. 2000) was also included for comparison. For α-AApeptide, α-AA4 is the most prominent one, with an IC_{50} of 38μM, which is comparable to the previously reported β-peptides and peptoids (Knight et al. 2002; Kritzer et al. 2004; Hara et al. 2006), and is only 4-5 fold less potent than the reported p53-drived wild type peptide (Garcia-Echeverria et al. 2000). Consistent to previous reports (Kussie et al. 1996), the preliminary structure and activity relationship (SAR) study here suggests that the inclusion of functionalities of the three key amino acids "Phe, Typ, and Leu" are important to maintain the strong binding affinity, which are present in all sequences but α-AA1. Compared to α-AA2, the change of Leu to Val in α-AA1 decreases the binding affinity to at least 10-fold. Further, that observation of α-AA4's much higher activity than α-AA2 indicates that better activities are possessed by longer sequences. The longer sequence may possibly have a better stabilized backbone conformation.

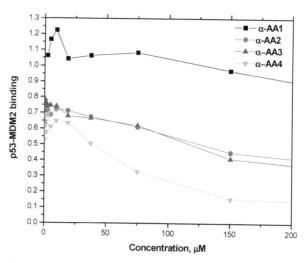

Fig. 7. Plots of ELISA assay for the inhibition of p53-MDM2 interaction by α-AApeptides Figure is adapted from (Hu et al. 2011).

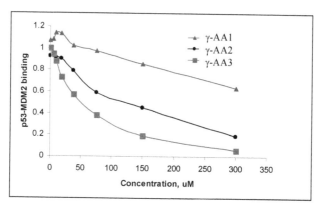

Fig. 8. Plots of ELISA assay for the inhibition of p53-MDM2 interaction by γ-AApeptides. Figure is adapted from (Niu et al. 2011).

AApeptides	IC$_{50}$ (μM)
α-AA1	>1000
α-AA2	120 ± 10
α-AA3	120 ± 16
α-AA4	38 ± 8
γ-AA1	> 400
γ-AA2	120 ± 15
γ-AA3	50 ± 8
p53-derived peptide (Ac-QETFSDLWKLLP)	8.7

Table 1. ELISA results of AApeptides for the disruption of p53/MDM2.

Finally, since α-**AA4** differs from α-**AA3** in only one residue, the side chains not involved in the recognition of MDM2 may also play a substantial role in the binding event. In this case, the Phe side chain of α-**AA3** may be either in the hydrophobic binding cleft, or near the binding domain, clashing with the residues of MDM2 and thereby raising up their binding energy. For γ-AApeptides, γ-**AA3** turns out to be the most effective inhibitor, with an IC_{50} value of 50μM, which is comparable to the most active α-AApeptide and is also only a few fold less active than p53-derived peptide (Garcia-Echeverria et al. 2000). Compared to the others, replacement of Phe by Leu in γ-**AA1** peptide results in a significant loss of activity. Whereas γ-**AA2** and γ-**AA3** both have the required "Phe, Typ, and Leu", the slight difference of their side chain functionalities between Phe and Leu results in more than two fold difference in activity, suggesting that even the side chains are also involved in the binding pocket. This observation is similar to the situation of α-AA peptides.

3.4 Computer modelling for bioactive AApeptides

The ELISA results were further confirmed by preliminary computer modeling studies (Figure 9 and 10), which shows that the side chains of Phe, Trp and Leu in the energy-minimized structures of both α-**AA4** and γ-**AA3** are able to overlap perfectly with those residues in the helical domain of natural peptide p53, indicating that α-**AA4** and γ-**AA3** should be able to mimic the recognition of p53 to MDM2 very well. Compared to γ-**AA3**, α-**AA4** appears to prefer an extended conformation when interacting with MDM2.

Fig. 9. Energy minimized (MM2) structures of α-**AA4** (green colored) and amino acids 17-29 of p53 helical domain (yellow colored). α-**AA4** is shown as sticks, and the three critical residues (Phe19, Trp23, and Leu 26) in p53 responsible for binding to MDM2 are also presented in sticks and colored in red. Figure is adapted from (Hu et al. 2011).

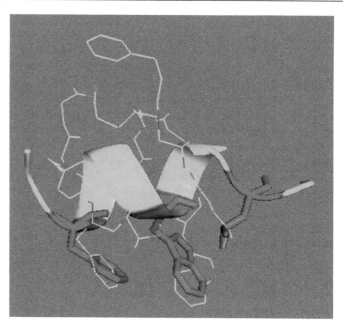

Fig. 10. Energy minimized (MM2) structures of **γ-AA3** (blue colored) is superimposed with the amino acids 17-29 of p53 helical domain (green colored). Three critical residues (Phe19, Trp23, and Leu 26) in p53 responsible for binding to MDM2 are presented in as sticks and colored in red. **γ-AA3** is shown as the wire frame presentation. Figure is adapted from (Niu et al. 2011).

3.5 Summary

Taken together, these results demonstrated AApeptides as a novel class of peptidomimetics. It is also noteworthy that both classes of AApeptides bear excellent selectivity, with different sequences giving different activities, instead of a random interaction with proteins. For example, α-AA4 is the strongest inhibitor among all the demonstrated α-AApeptides, while α-AA1 is a poor inhibitor, and α-AA2, α-AA3 are weak inhibitors. Similarly, γ-AA3 appears to be the best inhibitor, while γ-AA1 turns out to be the worst.

Detailed structure-activity relationship studies for AApeptides with various lengths and distribution of functional groups along the backbone are currently ongoing, which should provide valuable information for rational design of AApeptide library for drug discovery and chemical biology research. Generally, much more potent AApeptide derivatives are expected with the stabilization of secondary structure, introduction of halogen atoms, and computer modeling-aided design (Hara et al. 2006; Murray et al. 2007; Michel et al. 2009).

4. Stability of AApeptides

One significant advantage of peptidomimetics is the superior resistance to proteolysis, owing to their unnatural backbones. To find out the stability of our AApeptides in this regard, representative sequences of α- and γ- AApeptides (α-AA3, γ-AA3) were incubated with proteases at the concentration of 0.1 mg/mL in 100 mM pH 7.8 ammonium bicarbonate buffer for 24 hours. **α-AA3** was mixed with chymotrypsin, trypsin, and pronase, respectively; and **γ-AA3** was mixed with chymotrypsin, thermolysin, and pronase, respectively. All the reaction mixtures were then analyzed by HPLC. The retention time and integrations of eluted peaks were compared with those of peaks representative of the starting materials.

As shown in figure 11 and figure 12, whereas conventional peptides are susceptible to proteolysis, especially by chymotrypsin and pronase, both **α-AA3** and **γ-AA3** are highly resistant to enzymatic hydrolysis within 24 hours. There are, however, a small shoulder observed for both types of AApeptides at 37ºC with or even without incubation with proteases, which takes up less than 5% of the total volume and is presumably due to the isomerization of the syn/anti tertiary amide bonds in the peptide backbones.

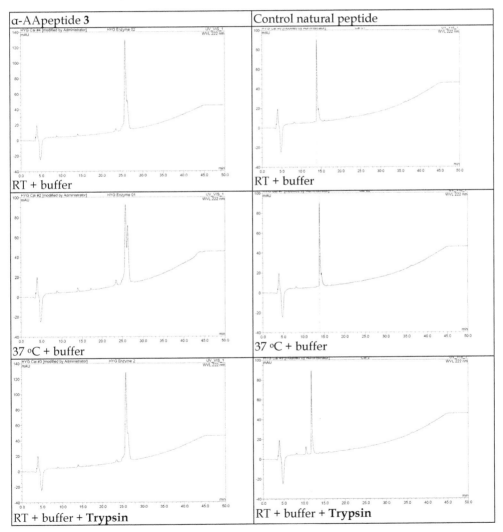

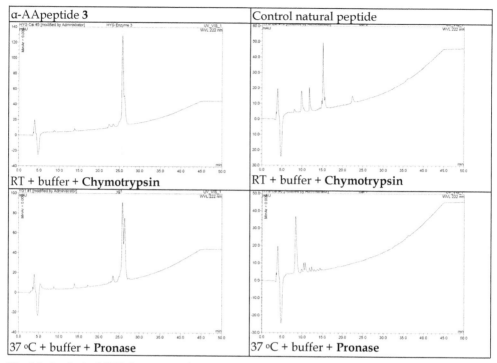

Fig. 11. Analytical HPLC spectra of **α-AA3** and control α-peptide after their incubations with different proteases. Figure is adapted from (Hu et al. 2011).

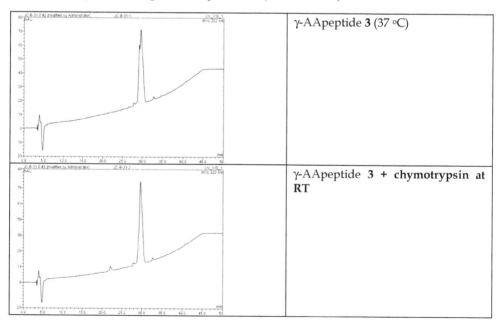

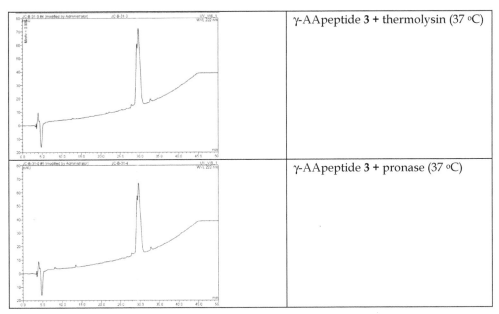

Fig. 12. Incubation of γ-AA3 with different proteases. Figure is adapted from (Niu et al. 2011).

5. Conclusions

In conclusion, we have developed a new class of peptidomimetics – the AApeptides that can have amino acid side chains at either α-, or γ-position. This family of peptides can be readily synthesized on solid phase by standard monomer building block approach, using α or γ-substituted N-acylated-N-Fmoc-amino ethyl amino acid building blocks. Given the availability of countless types of carboxylic acids, the AApeptides are amenable to potential derivatizations with a wide variety of side chains in a simple and straightforward manner, indicating its promising applications in library based drug screening. The preliminary results show that AApeptides possess significant bioactivities including the mimicry of p53 to successfully inhibit the p53/MDM2 protein-protein interaction, the selectivity in binding MDM2 protein, and the excellent stability towards enzymatic degradations. Hence, it is conceivable that a continuing development of sequence-specific AApeptides would enrich the current types of functional peptidomimetics, and expand the applications of peptide mimics in biomedical research including the modulation of protein-protein interactions. Future work will involve the systematic studies using X-ray crystallography, Circular Dichroism (CD), and 2D-NMR to understand the structure requirements of AApeptides to adopt predicted conformations, which will help the design of functional AApeptides. More specifically, the optimizations of AApeptide sequences to achieve a better inhibition of p53/MDM2 interaction as well as other carbohydrates/proteins/nucleic acids interactions are also urgent and are currently under investigation.

6. Acknowledgement

This work is supported by USF start-up fund (Cai) and NIH CA 118210 (Chen).

7. References

Alluri, P. G., M. M. Reddy, et al. (2003). "Isolation of protein ligands from large peptoid libraries." *J Am Chem Soc* 125(46): 13995-4004.

Arkin, M. R. and J. A. Wells (2004). "Small-molecule inhibitors of protein-protein interactions: progressing towards the dream." *Nat Rev Drug Discov* 3(4): 301-17.

Arndt, H. D., B. Ziemer, et al. (2004). "Folding propensity of cyclohexylether-delta-peptides." *Org Lett* 6(19): 3269-72.

Balint, E. E. and K. H. Vousden (2001). "Activation and activities of the p53 tumour suppressor protein." *Br J Cancer* 85(12): 1813-23.

Bautista, A. D., J. S. Appelbaum, et al. (2010). "Bridged beta(3)-peptide inhibitors of p53-hDM2 complexation: correlation between affinity and cell permeability." *J Am Chem Soc* 132(9): 2904-6.

Boeijen, A., J. van Ameijde, et al. (2001). "Solid-phase synthesis of oligourea peptidomimetics employing the Fmoc protection strategy." *Journal of Organic Chemistry* 66(25): 8454-8462.

Brooks, C. L. and W. Gu (2006). "p53 ubiquitination: Mdm2 and beyond." *Mol Cell* 21(3): 307-15.

Cheng, R. P., S. H. Gellman, et al. (2001). "beta-Peptides: from structure to function." *Chem Rev* 101(10): 3219-32.

De Vincenzo, R., G. Scambia, et al. (1995). "Effect of synthetic and naturally occurring chalcones on ovarian cancer cell growth: structure-activity relationships." *Anticancer Drug Des* 10(6): 481-90.

Debaene, F., J. A. Da Silva, et al. (2007). "Expanding the scope of PNA-encoded libraries: divergent synthesis of libraries targeting cysteine, serine and metallo-proteases as well as tyrosine phosphatases." *Tetrahedron* 63(28): 6577-6586.

Dervan, P. B. (1986). "Design of sequence-specific DNA-binding molecules." *Science* 232(4749): 464-71.

Ding, K., Y. Lu, et al. (2005). "Structure-based design of potent non-peptide MDM2 inhibitors." *J Am Chem Soc* 127(29): 10130-1.

Dragulescu-Andrasi, A., S. Rapireddy, et al. (2006). "A simple gamma-backbone modification preorganizes peptide nucleic acid into a helical structure." *J Am Chem Soc* 128(31): 10258-10267.

Duncan, S. J., M. A. Cooper, et al. (2003). "Binding of an inhibitor of the p53/MDM2 interaction to MDM2." *Chem Commun (Camb)*(3): 316-7.

Galatin, P. S. and D. J. Abraham (2004). "A nonpeptidic sulfonamide inhibits the p53-mdm2 interaction and activates p53-dependent transcription in mdm2-overexpressing cells." *J Med Chem* 47(17): 4163-5.

Garcia-Echeverria, C., P. Chene, et al. (2000). "Discovery of Potent Antagonists of the Interaction between Human Double Minute 2 and Tumor Suppressor p53." *J Med Chem* 43(17): 3205-3208.

Gellman, S. (2009). "Structure and Function in Peptidic Foldamers." *Biopolymers* 92(4): 293-293.

Goodman, C. M., S. Choi, et al. (2007). "Foldamers as versatile frameworks for the design and evolution of function." *Nat Chem Biol* 3(5): 252-62.

Grasberger, B. L., T. Lu, et al. (2005). "Discovery and cocrystal structure of benzodiazepinedione HDM2 antagonists that activate p53 in cells." *J Med Chem* 48(4): 909-12.

Graybill, T. L., M. J. Ross, et al. (1992). "Synthesis and Evaluation of Azapeptide-Derived Inhibitors of Serine and Cysteine Proteases." *Bioorganic & Medicinal Chemistry Letters* 2(11): 1375-1380.

Hara, T., S. R. Durell, et al. (2006). "Probing the structural requirements of peptoids that inhibit HDM2-p53 interactions." *J Am Chem Soc* 128(6): 1995-2004.

Hardcastle, I. R., S. U. Ahmed, et al. (2006). "Small-molecule inhibitors of the MDM2-p53 protein-protein interaction based on an isoindolinone scaffold." *J Med Chem* 49(21): 6209-21.

Haupt, Y., R. Maya, et al. (1997). "Mdm2 promotes the rapid degradation of p53." *Nature* 387(6630): 296-9.

Hayashi, R., D. Wang, et al. (2009). "N-acylpolyamine inhibitors of HDM2 and HDMX binding to p53." *Bioorg Med Chem* 17(23): 7884-93.

Horne, W. S. and S. H. Gellman (2008). "Foldamers with heterogeneous backbones." *Acc Chem Res* 41(10): 1399-408.

Horne, W. S., L. M. Johnson, et al. (2009). "Structural and biological mimicry of protein surface recognition by alpha/beta-peptide foldamers." *Proc Natl Acad Sci U S A* 106(35): 14751-6.

Hu, Y., X. Li, et al. (2011). "Design and synthesis of AApeptides: a new class of peptide mimics." *Bioorg Med Chem Lett* 21(5): 1469-71.

Jimenez, G. S., S. H. Khan, et al. (1999). "p53 regulation by post-translational modification and nuclear retention in response to diverse stresses." *Oncogene* 18(53): 7656-65.

Knight, S. M., N. Umezawa, et al. (2002). "A fluorescence polarization assay for the identification of inhibitors of the p53-DM2 protein-protein interaction." *Anal Biochem* 300(2): 230-6.

Kritzer, J. A., J. D. Lear, et al. (2004). "Helical beta-peptide inhibitors of the p53-hDM2 interaction." *J Am Chem Soc* 126(31): 9468-9.

Kritzer, J. A., O. M. Stephens, et al. (2005). "beta-Peptides as inhibitors of protein-protein interactions." *Bioorg Med Chem* 13(1): 11-6.

Kubbutat, M. H., S. N. Jones, et al. (1997). "Regulation of p53 stability by Mdm2." *Nature* 387(6630): 299-303.

Kumbhani, D. J., G. V. Sharma, et al. (2006). "Fascicular conduction disturbances after coronary artery bypass surgery: a review with a meta-analysis of their long-term significance." *J Card Surg* 21(4): 428-34.

Kussie, P. H., S. Gorina, et al. (1996). "Structure of the MDM2 oncoprotein bound to the p53 tumor suppressor transactivation domain." *Science* 274(5289): 948-53.

Lee, H. J., J. W. Song, et al. (2002). "A theoretical study of conformational properties of N-methyl azapeptide derivatives." *Journal of the American Chemical Society* 124(40): 11881-11893.

Li, X., Y. D. Wu, et al. (2008). "Alpha-aminoxy acids: new possibilities from foldamers to anion receptors and channels." *Acc Chem Res* 41(10): 1428-38.

Liu, Y. and M. Kulesz-Martin (2001). "p53 protein at the hub of cellular DNA damage response pathways through sequence-specific and non-sequence-specific DNA binding." *Carcinogenesis* 22(6): 851-60.

Lowe, S. W., H. E. Ruley, et al. (1993). "p53-dependent apoptosis modulates the cytotoxicity of anticancer agents." *Cell* 74(6): 957-67.

McLure, K. G., M. Takagi, et al. (2004). "NAD+ modulates p53 DNA binding specificity and function." *Mol Cell Biol* 24(22): 9958-67.

Michel, J., E. A. Harker, et al. (2009). "In Silico Improvement of beta3-peptide inhibitors of p53 x hDM2 and p53 x hDMX." *J Am Chem Soc* 131(18): 6356-7.

Momand, J., D. Jung, et al. (1998). "The MDM2 gene amplification database." *Nucleic Acids Res* 26(15): 3453-9.

Momand, J., G. P. Zambetti, et al. (1992). "The mdm-2 oncogene product forms a complex with the p53 protein and inhibits p53-mediated transactivation." *Cell* 69(7): 1237-45.

Murray, J. K. and S. H. Gellman (2007). "Targeting protein-protein interactions: lessons from p53/MDM2." *Biopolymers* 88(5): 657-86.

Nelson, J. C., J. G. Saven, et al. (1997). "Solvophobically driven folding of nonbiological oligomers." *Science* 277(5333): 1793-6.

Niu, Y., Y. Hu, et al. (2011). "[gamma]-AApeptides: design, synthesis and evaluation." *New J Chem* 35(3): 542-545.

Oliner, J. D., K. W. Kinzler, et al. (1992). "Amplification of a gene encoding a p53-associated protein in human sarcomas." *Nature* 358(6381): 80-3.

Oren, M. (1999). "Regulation of the p53 tumor suppressor protein." *J Biol Chem* 274(51): 36031-4.

Patch, J. A. and A. E. Barron (2002). "Mimicry of bioactive peptides via non-natural, sequence-specific peptidomimetic oligomers." *Curr Opin Chem Biol* 6(6): 872-7.

Pellegata, N. S., R. J. Antoniono, et al. (1996). "DNA damage and p53-mediated cell cycle arrest: a reevaluation." *Proc Natl Acad Sci U S A* 93(26): 15209-14.

Risseeuw, M. D., J. Mazurek, et al. (2007). "Synthesis of alkylated sugar amino acids: conformationally restricted L-Xaa-L-Ser/Thr mimics." *Org Biomol Chem* 5(14): 2311-4.

Robinson, J. A. (2008). "Beta-hairpin peptidomimetics: design, structures and biological activities." *Acc Chem Res* 41(10): 1278-88.

Seebach, D., P. E. Ciceri, et al. (1996). "Probing the helical secondary structure of short-chain beta-peptides." *Helvetica Chimica Acta* 79(8): 2043-2066.

Simon, R. J., R. S. Kania, et al. (1992). "Peptoids: a modular approach to drug discovery." *Proc Natl Acad Sci U S A* 89(20): 9367-71.

Stoll, R., C. Renner, et al. (2001). "Chalcone derivatives antagonize interactions between the human oncoprotein MDM2 and p53." *Biochemistry* 40(2): 336-44.

Trabocchi, A., F. Guarna, et al. (2005). "gamma- and delta-amino acids: Synthetic strategies and relevant applications." *Current Organic Chemistry* 9(12): 1127-1153.

Tsukamoto, S., T. Yoshida, et al. (2006). "Hexylitaconic acid: a new inhibitor of p53-HDM2 interaction isolated from a marine-derived fungus, Arthrinium sp." *Bioorg Med Chem Lett* 16(1): 69-71.

Tuwalska, D., J. Sienkiewicz, et al. (2008). "Synthesis and conformational analysis of methyl 3-amino-2,3-dideoxyhexopyranosiduronic acids, new sugar amino acids, and their diglycotides." *Carbohydr Res* 343(7): 1142-52.

Vassilev, L. T., B. T. Vu, et al. (2004). "In vivo activation of the p53 pathway by small-molecule antagonists of MDM2." *Science* 303(5659): 844-8.

Violette, A., M. C. Petit, et al. (2005). "Oligourea foldamers as antimicrobial peptidomimetics." *Biopolymers* 80(4): 516-516.

Whitty, A. and G. Kumaravel (2006). "Between a rock and a hard place?" *Nat Chem Biol* 2(3): 112-8.

Winssinger, N., R. Damoiseaux, et al. (2004). "PNA-Encoded Protease Substrate Microarrays." *Chem Biol* 11(10): 1351-1360.

Wu, Y.-D. and S. Gellman (2008). "Peptidomimetics." *Accounts of Chemical Research* 41(10): 1231-1232.

Zhao, J., M. Wang, et al. (2002). "The initial evaluation of non-peptidic small-molecule HDM2 inhibitors based on p53-HDM2 complex structure." *Cancer Lett* 183(1): 69-77.

7

Characterization of Protein-Protein Interactions via Static and Dynamic Light Scattering

Daniel Some and Sophia Kenrick
Wyatt Technology Corp.
USA

1. Introduction

Light scattering in its various flavors constitutes a label-free, non-destructive probe of macromolecular interactions in solution, providing a direct indication of the formation or dissociation of complexes by measuring changes in the average molar mass or molecular radius as a function of solution composition and time. It is a first-principles technique, thoroughly grounded in thermodynamics, permitting quantitative analysis of key properties such as stoichiometry, equilibrium association constants, and reaction rate parameters.

In the past, light scattering experiments on interacting protein solutions have been labor intensive and tedious, requiring large volumes of sample, and hence impeding widespread adoption by protein researchers. Recent advances in instrumentation and technique hold the promise for simplifying and automating measurements, as well as reducing sample requirements, thus broadening the appeal of these methods to the wider community of analytical biochemistry, biophysics, and molecular biology research. Pioneering work in automating and applying these measurements to equilibrium protein-protein interactions appeared in 2005-2006 (Attri & Minton, 2005a, 2005b; Kameyama & Minton, 2006). This chapter deals primarily with such automated methods.

2. Theory of light scattering from biomacromolecules in solution

2.1 Static light scattering

Static light scattering (SLS) measurements quantify the "excess Rayleigh ratio" R, which describes the fraction of incident light scattered by the macromolecules per unit volume of solution. Knowledge of R vs. scattering angle (θ) and concentration c may determine molar mass, size and self-interactions of the sample, while $R(t)$ will describe the kinetics of self-association or dissociation, via time-dependent changes in the average molar mass or size. Likewise, characterization of R vs. the composition ([A], [B], ...) of a multi-component system, such as hetero-associating proteins, may determine the stoichiometry and equilibrium binding affinity of such a system, as well as binding or dissociation kinetics.

The basic theory of static light scattering from macromolecules in solution is available in myriad publications, including elementary textbooks dealing with physical chemistry (e.g., van Holde et al., 1998; Teraoka, 2002) or essential references on polymers (e.g., Young, 1981). We will cite from these without further reference. More rigorous publications, particularly those dealing with non-ideal, multi-component solutions are found in the scientific literature (e.g., Blanco et al., 2011, and references therein).

2.1.1 Static light scattering in the ideal limit

Macromolecules in solution are subject to correlations arising from intermolecular potentials, which in turn affect the magnitude of scattered light. However, if the particles are few and far between, and the potentials between them sufficiently short-ranged, these correlations may be ignored, leading to what is known as "the ideal limit": essentially, the particles behave like an ideal ensemble of point particles.

The simplest picture of scattering from proteins in solution invokes the ideal limit, i.e., point-like particles with no interactions, much like the more commonly known ideal gas law for pressure and temperature. In this case, the scattering from particles much smaller than the wavelength of incident light, with detectors placed in the plane perpendicular to the incident polarization, can be described by Eq. (1):

$$R = \frac{4\pi^2 n_0^2}{N_A \lambda_0^4} \left(\frac{dn}{dc} \right)^2 Mc = K * Mc \tag{1}$$

In Eq. (1), N_A represents Avogadro's number, dn/dc is the protein's refractive increment, M the protein's molar mass, n_0 the solvent refractive index, λ_0 the wavelength in vacuum, and c the protein concentration in units of mass/volume. K^* incorporates the constants n_0, N_A, λ_0 and dn/dc.

The protein refractive increment dn/dc describes the change in refractive index of a solution relative to pure buffer, due to a mass/volume protein concentration c; this parameter may be readily measured by means of a common instrument known as a differential refractometer and is, fortuitously, nearly invariable for most proteins in standard aqueous buffers at any given wavelength (dn/dc=0.187 mL/g at λ=660 nm). High concentrations of excipient will affect dn/dc; adding for example arginine, which has a refractive index higher than that of most proteins, can even reduce dn/dc to zero such that no scattering occurs!

For a solution consisting of multiple macromolecular species, e.g., monomer + oligomers or A+B+AB complex, the total light scattered is the sum of intensities scattered by each species:

$$R = K * \sum_i M_i c_i = K * M_w c \tag{2}$$

Here M_i and c_i refer to the molar mass and concentration of each species i, M_w is the weight-averaged molar mass and c the total protein concentrations. We have assumed that all species have the same refractive increment and non-ideality may be ignored.

Upon inspection of Eq. (2) it becomes clear that given knowledge of the measurement conditions (solution refractive index, scattering wavelength), sample parameters (dn/dc), sample concentration (e.g., by means of a UV absorption or differential refractive index concentration detector), and excess Rayleigh ratio R, it is possible to determine the weight-averaged molar mass of macromolecules in the solution. If the solution is monodisperse (as is often the case in the course of chromatographic fractionation), then the molar mass of the solvated macromolecule may be determined.

Eq. (2) contains the reason that light scattering is famously sensitive to small quantities of dust or other particulates: if the mass if the dust particle is a million times that of the protein, only one-millionth the concentration of dust particles produces the same scattering intensity as the protein.

Generalization of Eq. (2) to species with different refractive increments is obvious, but for reasons of simplicity we will assume henceforth equal *dn/dc* for all proteins. **This "ideal gas law for light scattering" is generally applicable to characterization of specific protein-protein binding with equilibrium dissociation constants $K_D \leq 1$-10 μM.**

Angular dependence of the scattered intensity comes into play for larger particles such as protein aggregates whose radii exceed $\sim\lambda/50$. In the limit of transparent particles with radii below ~40 nm, this dependence is described by the Rayleigh-Ganz-Debye (RGD) equation:

$$R(\theta) = K * McP(\theta); \quad P(\theta) = 1 - \frac{16\pi^2}{3\lambda_0^2}\langle r_g^2 \rangle \sin^2(\theta/2) + \dots \tag{3}$$

For this reason SLS is often referred to as multi-angle light scattering (MALS). Here θ is the angle between the incident and scattered light rays within the plane perpendicular to the incident polarization, r_g is the radius of gyration, and higher order terms in $P(\theta)$ have been ignored. For globular proteins, r_g will be $\sim80\%$ the average geometrical radius. When $r_g < 8$-12 nm, $P(\theta) \sim 1$, angular dependence is eliminated, and the molecules are considered isotropic scatterers. Since the dimensions of most proteins and complexes are below 20 nm, for the remainder of this chapter we will assume isotropic scattering, i.e., $P(\theta) = 1$.

2.1.2 Analysis of protein complexes via static light scattering coupled to online chromatographic separation

Analytical size-exclusion chromatography (SEC) is often an unreliable measure of molar mass, particularly if the standards used to calibrate column elution times do not represent the sample well in terms of shape or column interactions. Because it does not need to make any assumptions regarding separation models or column calibration standards, flow-mode MALS is an invaluable extension of analytical SEC (SEC-MALS) or asymmetric-flow field flow fractionation (AF4-MALS). The analysis almost invariably occurs at concentrations well below 1 mg/mL, low enough to fall squarely within the ideal limit. Figure 1 shows a typical SEC-MALS experimental layout, combining a MALS detector with concentration analysis by means of UV/Vis absorption or differential refractometry (dRI).

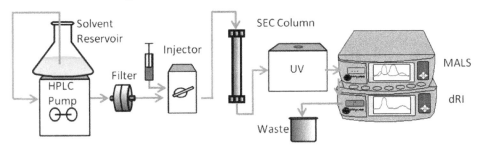

Fig. 1. SEC-MALS instrumentation.

While normally applied to characterize polydisperse ensembles, irreversible oligomers or other tightly-bound complexes, SEC-MALS may be used to assess reversible protein interactions, especially self-association (Bajaj et al., 2007). In this approach, the excess Rayleigh ratio and concentration is measured at multiple points along an eluting peak, and these data are fit to equations representing mass conservation, mass action and ideal light scattering identical to those in Section 2.1.3. Since the ratio of monomer to oligomer in a reversibly associating system is concentration-dependent, the change in weight-average molar mass across the peak should indicate dissociation of oligomers upon dilution in the column. The measurement may be repeated at different initial sample concentrations to enhance the analysis and establish whether the dissociation kinetics are fast or slow.

While this analysis can provide a good semi-quantitative characterization of reversible association, it is subject to certain systematic errors. The analysis must assume either rapid or very slow equilibration. As the sample proceeds through the column and detectors, it dilutes continuously; if the equilibration is neither very fast nor very slow compared to the elution time, the ratio of complex to monomer will not represent equilibrium conditions. Also, band-broadening between detectors means that the concentration measured in the UV or dRI concentration detector is somewhat different than that in the MALS cell, hence systematic errors in the analysis will arise. An integrated UV-SLS cell can eliminate the latter source of error (Bajaj et. al., 2004).

The advantages of analyzing reversible complexes via SEC-MALS are low sample quantity and clean data with little noise due to particulates if the size exclusion column and HPLC system are very clean. The column will separate any dust or aggregates from the sample.

2.1.3 Quantifying specific, reversible protein-protein binding via composition-gradient static light scattering

The use of stop-flow injections with well-defined concentrations permits true equilibrium analysis. In some instances the kinetics of association or dissociation may be analyzed as well. Composition-gradient light scattering apparatus, described in section 3, is more generally useful than SEC-MALS for studying protein-protein interactions. This section presents the principles of this approach.

The analysis of specific, reversible protein-protein interactions in the ideal limit, via light scattering measurements from a series of compositions, has been presented concisely by Attri & Minton (Attri & Minton, 2005b). The equations, with a minor change in notation, include the ideal light scattering law (4), mass action (5) and conservation of mass (6), shown below assuming up to two constituent monomeric species A and B:

$$\frac{R}{K^*} = \sum_{i,j} (iM_A + jM_B)^2 \left[A_i B_j \right] \tag{4}$$

$$K_{ij} = \frac{\left[A_i B_j \right]}{\left[A \right]^i \left[B \right]^j} \tag{5}$$

$$[A]_{total} = \sum_{i,j} i \left[A_i B_j \right], \quad [B]_{total} = \sum_{i,j} j \left[A_i B_j \right] \tag{6}$$

M_A and M_B represent the molar masses of constituent monomers A and B, respectively; i and j represent the stoichiometric numbers of A and B in the A_iB_j complex, with A_1B_0 and A_0B_1 representing the monomers of A and B; $[A_iB_j]$ represents the molar concentration of the A_iB_j complex; $[A]_{total}$ and $[B]_{total}$ represent the total molar concentration of A and B in solution; and $K_{i,j}$ is the equilibrium association constant relating equilibrium molar concentrations of the A_iB_j complex and free monomer. Light scattering and concentration data acquired over a series of compositions — multiple concentrations, in the case of self-association of a single species, or a series of A:B composition ratios in the case of hetero-association — are fit to Eq. (4) by means of a standard least-squares nonlinear curve fitting procedure. This technique is known as composition-gradient multi-angle static light scattering, CG-MALS or CG-SLS. Beyond the usual curve fitting algorithms, there is an added complication of solving first at each composition and fitted parameter iteration, the nonlinear system of Eqs. (5) (one for each complex present in equilibrium with monomers) + Eqs. (6). Examples of the system of equations to be fit are presented in Section 4.4.1.

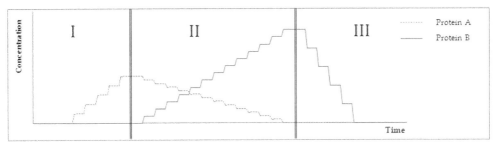

Fig. 2. Composition gradient for characterization of self- and hetero-association.

A typical composition gradient for the analysis of combined self-association and hetero-association, shown in Figure 2, includes three segments: I and III are concentration series in A and B individually to determine any self-association, while II is a "crossover gradient" stepping through a series of A:B composition ratios to characterize hetero-association. Figure 3a simulates LS signals for homodimer and homotrimer association, showing how the appropriate association model is well-determined by LS.

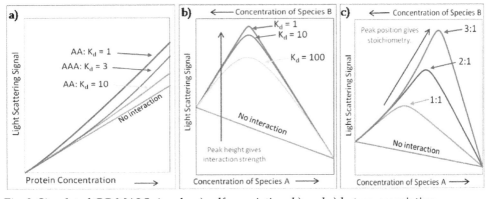

Fig. 3. Simulated CG-MALS signals. a) self association; b) and c) hetero-association.

The crossover gradient, though perhaps not intuitive to those accustomed to sigmoidal titration curves, is in fact quite efficient for analyzing stoichiometry and binding affinity even in the presence of a complex interaction that may include simultaneous self- and hetero-association. Figure 3b and Figure 3c depict a qualitative interpretation of the behavior of the light scattering signal for such a gradient: the position of the peak along the composition gradient axis indicates the stoichiometric ratio of the complex, while the height and shape of the peak indicate both the binding affinity and the true stoichiometry, i.e., discrimination between the formation of 1:1 or 2:2 complexes.

2.1.4 Non-ideal static light scattering for characterization of nonspecific protein interactions

Nonspecific protein interactions arise from various sources such as hard-core molecular repulsion, net charge, dipoles, hydrophobic patches, van der Waals interactions, hydration forces, etc. These are generally weak in relation to specific binding and so only become important at concentrations exceeding about 1 mg/mL. In contrast to site-specific lock-and-key binding, nonspecific interactions do not exhibit well-defined stoichiometry and do not generally saturate at some sufficiently high concentration. The dominant interaction may vary from attractive to repulsive or vice-versa as the concentration increases. The ill-defined, multi-sourced nature of nonspecific interactions lends itself to thermodynamic analysis in terms of small deviations from the ideal case described by Eq. (1), hence the designation "thermodynamic non-ideality." These interactions are of interest since they generally correlate to solubility, viscosity, and propensity for aggregation, and are also key to understanding crowded biomolecular environments such as the intracellular environment.

Akin to the virial expansion of the osmotic pressure as a power series in concentration, light scattering of dilute solutions may be analyzed in terms of a virial expansion, which actually uses the same pre-factors (virial coefficients A_2, A_3, or B_{22}, B_{222}, etc.) as the osmotic pressure equation though with a slightly different functional form:

$$\frac{R}{K^*} = \frac{Mc}{1 + 2A_2Mc + 3A_3Mc^2 + \ldots} \quad or \quad \frac{K^*c}{R} = \frac{1}{M} + 2A_2c + 3A_3c^2 + \ldots \qquad (7)$$

In many cases a protein's nonspecific interactions may be modeled as those of hard, impenetrable spheres, albeit with an effective specific volume v_{eff} different from that of the molecule's actual specific volume, resulting in a virial expansion containing only one independent parameter describing the non-ideality (Minton & Edelhoch, 1982):

$$\frac{R}{K^*} = \frac{Mc}{1 + 8v_{eff}c + 30\left(v_{eff}c\right)^2 + \ldots} \qquad (8)$$

Each virial coefficient may be expressed in terms of the effective volume. As the lowest–order correction to ideal light scattering, $A_2 = 4v_{eff}/M$ tends to be of greatest interest for characterizing nonspecific interactions. Unscreened, long-range charge-charge interactions are not fit well by the effective hard sphere model.

An interesting and counterintuitive feature of purely (or primarily) repulsive interactions is that the LS intensity is not monotonic with concentration. Rather it initially rises as expected from the ideal LS equation, then plateaus, and eventually declines with higher concentration (see high-concentration case study, Section 5.5). Many non-associating proteins exhibit light scattering behavior which is fit well by the effective hard sphere assumption including scattering that eventually decreases with concentration (Fernández & Minton, 2008). The same work described CG-SLS apparatus suitable for high concentrations.

2.1.5 Quantifying repulsive and weakly attractive protein interactions via composition-gradient static light scattering at high concentration

Attractive nonspecific interactions, though weak, will at high enough concentrations invariably lead to the formation of transient clusters which can be analyzed in terms of specific reversible oligomerization, rather than in terms of virial coefficients. The same is not true of repulsive interactions. Hence it is possible to segregate the repulsive interactions into the virial coefficients and treat the attractive interactions as specific self-association.

The analysis is further simplified by assuming not only that the monomers and oligomers behave as effective hard spheres but also that all species have the same effective specific volume v_{eff}. An algorithm has been developed to include the effect of varying thermodynamic activity on the apparent association constants (Minton, 2007). This approach has been shown to work well for enzymes (Fernández & Minton, 2009) as well as antibodies (Scherer et al., 2010). While not quite as rigorous, we have found that a reduced analysis based on Eq. (9) serves to reproduce the essential behavior at high protein concentration in terms of self-associating oligomers subject to an effective hard sphere repulsion:

$$\frac{R}{K^*} = \frac{\sum_i iMc_i}{1 + 8v_{eff}c + 30\left(v_{eff}c\right)^2 + \ldots} \tag{9}$$

2.2 Dynamic light scattering

Rather than measuring the time-averaged intensity of scattered light, dynamic light scattering (DLS) measures the intensity fluctuations which arise from Brownian motion of the scattering molecules. The fluctuations are mathematically processed to produce an autocorrelation spectrum, which is then fit to appropriate functional forms to assess the translational diffusion constant D_t. D_t can be related via the Stokes-Einstein equation to a characteristic dimension, the hydrodynamic radius r_h, which is just the radius of a sphere with the diffusion constant D_t. The theory of DLS is covered in myriad sources, from textbooks (e.g., Teraoka, 2002) to peer-reviewed scientific literature.

DLS has certain practical advantages over SLS. In particular, DLS enjoys a relative immunity to stray light, which permits robust measurements in small volumes with free surfaces, such as a microwell plate. On the other hand, DLS is not as sensitive as SLS and so requires higher concentrations, limiting the range of binding affinities it can measure.

2.2.1 Analyzing protein-protein binding via composition-gradient dynamic light scattering in the ideal limit

The same nonspecific interactions leading to thermodynamic non-ideality in SLS do affect the diffusion constant and apparent r_h measured by DLS. In the ideal limit corresponding to a sufficiently dilute solution, this may be ignored. A solution consisting of multiple species

A_iB_j will exhibit a diffusion constant which is the z-average of the diffusion constants $D_{t,ij}$ of the individual species, leading to an average hydrodynamic radius as shown in Eq. (10):

$$\frac{1}{r_h^{avg}} = \sum_{i,j} \frac{1}{r_{h,ij}} M_{ij}^2 \left[A_i B_j \right] \bigg/ \sum_{i,j} M_{ij}^2 \left[A_i B_j \right] \tag{10}$$

Upon measuring a series of concentrations or compositions like that shown in Figure 2, one can determine stoichiometry and equilibrium association constants by analyzing the apparent diffusion constants in terms of Eq. (10) in combination with Eq. (5) and Eq. (6). This technique is known as composition-gradient dynamic light scattering (CG-DLS).

Unlike the molar mass measured by static light scattering, it is not obvious or straightforward to construct the hydrodynamic radius or diffusion constant of an associating complex, given the hydrodynamic radii of its constituent species. This is especially true for stoichiometries higher than 1:1 where the geometry could vary from compact to extended, leading to significantly different diffusion constants. Composition-gradient DLS data has been shown to successfully analyze binding of globular proteins to an A_iB_j complex by assuming a power law relationship (Hanlon et. al, 2010):

$$r_{h,ij} = \left(i r_{h,A}^\alpha + j r_{h,B}^\alpha \right)^{\frac{1}{\alpha}} \tag{11}$$

The best fits for different associating systems resulted in α values ranging from ~2.8 for homodimers and 1:1 complexes, to ~ 2.0 for a 2:1 stoichimetry. In this work, CG-DLS in a microwell plate reader provided important benefits, including low sample consumption and the ability to measure the same samples at multiple temperatures in order to obtain enthalpy and entropy of the interaction via van 't Hoff analysis.

2.2.2 Analyzing nonspecific interactions via dynamic light scattering

Nonspecific interactions give rise to thermodynamic non-ideality in DLS as well as MALS. The first-order correction to the translational diffusion constant incorporates a combination of parameters: the second osmotic virial coefficient A_2 (B_{22}), the specific volume of the molecule v_{sp}, and the first-order correction to the molecule's friction coefficient due to concentration ζ_1, as presented in Eq. (12) (Teraoka, 2002).

$$D_t = D_0 \left(1 + k_D c + \ldots \right); \quad k_D = 2A_2 - 2v_{sp} - \zeta_1 \tag{12}$$

We can expect v_{sp} to remain approximately constant for a given protein in different buffer systems, while ζ_1 actually includes additional A_2 dependence. A measurement of k_D for a series of monoclonal antibodies in different buffers exhibits excellent correlation with A_2 (Lehermayer et al., 2011).

The sample concentrations needed to measure k_D are comparable to those needed to measure A_2, but the volumes can be much smaller. Hence, in order to track trends in nonspecific interactions with buffer modifications such as pH or ionic strength, k_D (particularly as measured in a DLS plate reader) can be a low-volume, high-throughput surrogate for CG-MALS A_2 analysis. Unlike CG-MALS, however, currently the high-concentration behavior of CG-DLS is insufficiently understood to interpret data in the 10-200 mg/mL range.

3. Composition-gradient light scattering instrumentation

3.1 Detectors

Light scattering instrumentation for solutions has evolved considerably in the past two decades, resulting in systems that make characterization of protein interactions via CG-MALS and CG-DLS both feasible and accessible. Current top-of-the-line commercial MALS instruments provide a dynamic range covering protein solutions from tens of ng/mL up to hundreds of mg/mL (HELEOS, Wyatt Technology Corp.), accessing interactions with K_D from sub-nM to mM. Closed systems employing low-stray-light flow cells are important for low through moderately high concentrations, but are not suitable for the most concentrated protein samples that tend to be viscous. Easily-cleaned, microcuvette-based systems are better suited to the latter measurements.

A large selection of microcuvette-based DLS detectors is commercially available (Zetasizer series, Malvern Instruments; DynaPro series, Wyatt Technology; etc.). The lowest protein concentration range that these can analyze is on the order of 10-100 μg/mL, accessing an interaction range from tens of nM to tens of μM. Of particular note is the DynaPro DLS plate reader which can be integrated with standard liquid handling robotics to prepare automatically low-volume composition gradients.

Some instruments provide simultaneous SLS and DLS detection (HELEOS+QELS (flow cell or microcuvette), NanoStar (microcuvette), Wyatt Technology Corp; Zetasizer μV (flow cell or microcuvette), Malvern Instruments).

3.2 Automated composition-gradient delivery systems

An automated composition-delivery system for CG-MALS or CG-DLS is similar in many ways to standard stop-flow apparatus: two or more solutions are mixed by pumping via syringe pumps through a static mixer, an aliquot is delivered to the detector, and the flow stopped. The syringe pumps are operated at different flow ratios in order to obtain different compositions.

The most significant added requirement for light scattering is good in-line filtration of the solutions in order to eliminate large aggregates and particles generated by airborne dust, mechanical motion of syringes and valves, or protein films sloughing off surfaces. The pore size should be on the order of 0.1 μm or less. One approach (Attri & Minton, 2005a) is to add an in-line filter after the mixing point, illustrated in Figure 4. The key disadvantage of this

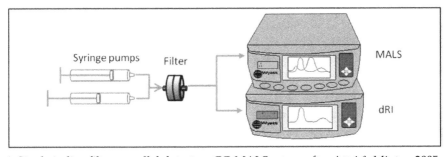

Fig. 4. Single in-line filter, parallel detectors CG-MALS setup, after Attri & Minton 2005a.

single-filter architecture is the changing chemical environment on the filter membrane: proteins will adhere to and release from the filter unpredictably as the environment changes. This is of particular concern in a tight-binding hetero-association analysis carried out at low protein concentrations. An in-line concentration detector is crucial.

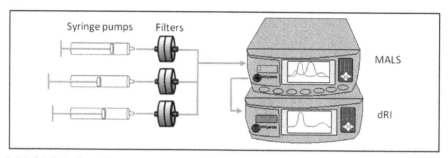

Fig. 5. Multiple in-line filters, serial detectors CG-MALS setup (Calypso, Wyatt Technology Corp.). UV or dRI concentration detector is optional.

Another approach is to flow each solution through a dedicated filter, as shown in Figure 5. In preparation for the gradient, each solution is pumped through its own filter and associated lines until saturation is reached so that, in the course of a subsequent composition gradient, well-defined compositions are produced reliably. This setup potentially eliminates the need for an in-line concentration detector if the stock protein solution concentrations are known prior to loading.

CG-MALS systems typically include light scattering and concentration detectors. The setup of Figure 5 shows a common approach, serially connected detectors, much like that of SEC-MALS. In order to achieve accurate correspondence between the concentration in the MALS flow cell and that in the concentration detector, both cells must be fully flushed with each composition. This can require relatively large sample volume. An alternative approach is to split the flow between the two detectors, as shown in Figure 4. Careful calibration of the flow resistance and delay between the two detectors is required to match the concentrations at the end of each injection. Additional care must be taken to ensure that laboratory temperature fluctuations, clogged capillaries or viscosity changes do not alter the split ratio between the detectors. The parallel detector configuration affords a smaller injection volume per composition.

Automation would not be complete without control and analysis software. Currently the only commercially available hardware/software package integrating syringe pump control, MALS data acquisition, and data analysis of equilibrium and kinetic macromolecular interactions is the Calypso CG-MALS system (Wyatt Technology, Santa Barbara).

4. Practical challenges of composition-gradient light scattering

4.1 Sample and buffer preparation

Due to the sensitivity of light scattering to the presence of just a few large particles, and especially in the absence of a separation step like SEC or FFF, particle-free solutions are essential in CG-MALS. Even though the composition-gradient apparatus provides in-line filtration, samples and solvents must be pre-filtered to the smallest practical pore size into very clean glassware or sterile, disposable containers. Solvents and buffers are generally filtered via bottle-top vacuum filters or large syringe-tip filters with pore sizes of 0.1-0.2 μm. Samples should be diluted to a bit above the appropriate working concentration in *filtered* solvent and then filtered with a syringe-tip filter to the smallest allowable pore size, most commonly 0.02-μm (e.g., Anotop filter, Whatman). All filters should be flushed to wash out any particles prior to introducing sample or final buffers.

4.2 Maintaining clean experimental apparatus

Regular cleaning and maintenance of the LS detectors and sample delivery apparatus are imperative for reliable, reproducible data. As a general rule, after each experiment, the instruments should be flushed with a buffer in which the sample is soluble before changing to storage or cleaning solutions. Common detergents for removing protein and polymer residue from glass and plastic surfaces include 5% v/v Contrad 70 and 1% w/v Tergazyme. Other methods useful for cleaning a dirty system include flushing with a high-salt (0.5-1.0 M NaCl) solution, 20-30% alcohol in water, or 10% nitric acid, as well as manual disassembly and cleaning. Salt and protein residues may be removed from syringes or valves by sonication.

Cleanliness of the instruments and buffers should be verified by observing noise in the MALS signals as the solutions flow through the system.

4.3 Designing optimal methods

The key parameters in CG-MALS experiment design are: 1) stock concentrations; 2) number and spacing of composition steps; 3) injected volume per step; and 4) equilibration time.

4.3.1 Determining optimal concentrations

Since molecular interactions are generally concentration-dependent, it is important to estimate the right concentration range that will, on the one hand, be high enough to produce a significant amount of complex, but on the other, be low enough so as not to saturate the complex leaving no free monomer. A general rule-of-thumb, assuming one has an estimate of K_D, is to prepare stock solution concentrations at 5-10x K_D. A more sophisticated approach is to perform a series of CG-MALS simulations assuming different concentrations, K_D values and even association schemes (e.g. 1:1 or 2:1), selecting composition gradients that best discriminate between reasonably feasible models.

For self-association, a concentration gradient should include concentrations low enough that essentially no self-association occurs, as shown at the low-concentration end of Figure 3a where the no-interaction signal coincides with the associating signal. The gradient should also include concentrations high enough that at least 20-30% of the LS intensity arises from oligomers, as shown on the high-concentration end of Figure 3a.

For heteroassociation, the optimal A:B stock concentration ratio is not necessarily the stoichiometric ratio, but depends on the molar masses of the molecular species. For good contrast, the total LS signal at 100% A (right side of Figure 3b) and 100% B (left side of Figure 3b) will be nearly equal, i.e., $M_A c_A \sim M_B c_B$ where c_A and c_B refer to the stock concentrations of A and B. This should be balanced against centering the LS peak close to the center of the crossover gradient. In particular, juggling these competing considerations can be tricky when the molar masses differ by a factor of 3 or more. If the mass ratio is large it may be better to perform a titration-like gradient where each injection includes a fixed concentration of the larger molecule, but varies the concentration of the smaller molecule.

Once the concentration ratio has been selected, the overall concentrations of A and B in the heteroassociation gradient should be chosen to discriminate well between K_D values within a reasonably expected range. For example, the conditions of Figure 3b discriminate well between K_D values of 1, 10 and 100, but would not be conclusive if the actual K_D is 0.1 or 1000.

An initial CG-MALS analysis may yield multiple association models that fit the data well. Simulations of light scattering from new composition gradients can assist in judiciously designing a follow-on experiment to refine the analysis by eliminating some of the first-round models. Such simulations are incorporated into the Calypso software.

The concentrations required to measure nonspecific interactions characterized by the second virial coefficient (A_2) typically range from 2-20 mg/mL. For proteins, an initial estimate of A_2 may be calculated by assuming a hard sphere of the same molecular weight (M) and hydrodynamic radius (r). The stock concentration needed to achieve a 15% contribution to the total scattered intensity from the A_2 term (see Eq. (7)) can be calculated as per Eq. (13):

$$c_{stock} = \frac{0.15}{2 A_2^{sphere} M} ; \quad A_2^{sphere} = \frac{16 \pi N_A r^3}{3 M^2} \tag{13}$$

4.3.2 Composition steps

An adequate number of compositions must be evaluated for proper fitting of CG-MALS/DLS data to an appropriate interaction model. For nonspecific interactions or simple homodimerization, at least five non-zero concentrations are recommended. Likewise, at least 5 composition steps are required for 1:1 binding. More complicated interactions forming larger numbers of species in equilibrium typically require 8-10 different compositions or more.

Sometimes the composition steps, rather than being distributed evenly across a gradient, can be focused in a specific region in order to make best use of the available sample, as shown in Figure 6.

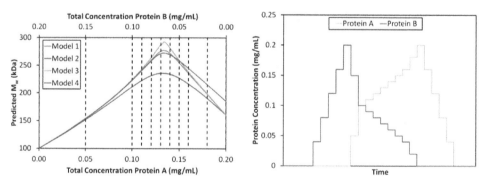

Fig. 6. Simulated interaction data for four possible interaction models (left) and corresponding CG-MALS method (right) focusing compositions around region of interest. Dashed vertical lines indicate plateau compositions.

4.3.3 Step volumes and equilibration time

The volume of sample introduced to the detectors at each composition must suffice to flush out completely the previous contents of the cell. At low sample concentration, or for particularly "sticky" samples, adsorption onto surfaces (especially the in-line filter of Figure 4) may necessitate increased step volumes. The required injection size may also vary with flow rate as well as detector configuration, and so should be determined experimentally for any set of conditions. The proper step volume may be assessed by running an ascending gradient followed by a descending gradient at a series of injection volumes: as the volume increases, the signals will match more closely.

After each injection, flow is stopped and the sample given time to reach equilibrium. Often the time scale for the reaction is faster than the dead time (the time between mixing and reaching the flow cell), but when this is not the case, ample time should be allotted after each injection for equilibration.

Where sample volume is scarce or where high concentration or viscosity prevents performing stop-flow experiments, CG-MALS analyses may be performed using a microcuvette. Stock solutions for each composition are prepared in advance. The light scattering intensity from each sample is measured using a calibrated cuvette and analyzed as for a flow system. Microcuvettes must be carefully cleaned and dried between samples.

4.4 Data analysis

CG-MALS data analysis protocols include two distinct segments: pre-processing and model fitting. The former comprises basic steps common to many measurements: baseline subtraction, application of proportionality (calibration or conversion) factors, smoothing, and selecting the data points for analysis. Specific to multi-angle light scattering are despiking and detector selection, since the main source of noise is foreign particles that primarily affect lower scattering angles and always generate positive signals. For equilibrium analysis, data should be selected after equilibration at each step, and usually a range of data points from each composition step are averaged to provide a single value from each detector.

4.4.1 Equilibrium models: Fitting and interpretation

The essential parameters in a CG-MALS model are monomer molar mass M_A and M_B; association stoichiometries ij; association constants $K_{A,ij}$; and incompetent (inactive) fractions $f_{incomp,A}$, $f_{incomp,B}$. The latter refer to protein molecules in the stock solution that are incompetent to participate in the interaction due to mutation, misfolding, aggregation, etc. In the course of fitting the data, a set of stoichiometries ij must be selected. Parameters such as monomer molar mass and incompetent fractions may be constrained to known values or floated to be optimized in the fit. Additionally, constraints may be imposed on the association constants, e.g., models of equivalent binding sites or isodesmic association confer specific relationships between the $K_{A,ij}$, as described in Table 1.

Standard iterative non-linear curve fitting algorithms, such as Levenberg-Marquardt, are implemented. For each composition, the total concentration of each monomer species is known either from precise dilution or by measurement with a UV or dRI detector. At each iteration of the free parameters, the equations for mass action and mass conservation (Eqs.

(5) and (6)) are solved; then the light scattering is computed (Eq. (4)) and compared to the measured value, the difference thereof serving as the minimization function. The result of fitting the data to a particular model will provide association constants plus any other free parameters, as well as a measure of goodness of fit, such as χ^2.

A broad range of useful equilibrium association models may be implemented, including any combination of self-interactions (formation of oligomers) and hetero-associations (stoichiometries of 1:1, 2:2, 1:n, etc.). Several common association models for proteins are presented in Table 1. Examples of these and more complex association schemes are discussed in Section 5.

Although useful, appropriate fitting of CG-MALS data does not require *a priori* knowledge of the interaction stoichiometry or system constraints. For a well-designed experiment, the best fit of the data should naturally converge on a single solution. This is illustrated in Figure 7 where incorrect models are applied to LS data for 1:1 and 1:2 interactions. In Figure 7A, which depicts a 1:2 interaction with equivalent binding sites as for antibody-antigen binding, a first guess of 1:1 interaction creates a fitted curve that does not peak at the correct stoichiometric ratio and clearly does not fit the data. Similarly, applying combined 1:1 and 1:2 stoichiometries, unconstrained for equivalent and independent binding sites, to a system with only one binding site results in the fitting algorithm eliminating the contribution from the 1:2 species (LS contribution from $AB_2 = 0$ for all compositions, Figure 7B).

In many instances, the expected interaction scheme fits the data well, resulting in low χ^2 and random residuals. Otherwise different stoichiometric models should be tested until the measured LS behavior is well matched. If the experiment design was far from optimal for the true system behavior or the interaction is particularly complex, more than one model may fit the data equally well. Several strategies may be brought to bear on selecting the most appropriate scheme, including Occam's razor (i.e., the simplest model that fits the data) and information from other techniques such as crystallographic structure or NMR analysis of binding sites. Simulation tools are useful in designing follow-up experiments to discriminate between multiple possibilities.

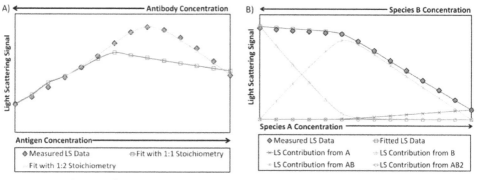

Fig. 7. Proper fitting of CG-MALS data requires the correct association model. A) Fitting of 1:2 interaction by 1:1 or 1:2 stoichiometry. B) Best fit of interaction between chymotrypsin (A) and bovine pancreatic trypsin inhibitor (B) from crossover gradient in Section 5.2.1 requires only 1:1 interaction.

Interaction Model	Unknowns	Governing Equations
Dimerization (e.g., Section 5.1.1)	$K_{20} = \frac{[A_2]}{[A]^2}$	$R/K^* = M_A^2\left([A] + 4K_{20}[A]^2\right)$ $[A]_{total} = [A] + 2K_{20}[A]^2$
1:1 hetero-association (e.g., Section 5.2.1)	$K_{11} = \frac{[AB]}{[A][B]}$	$R/K^* = M_A^2[A] + M_B^2[B] + (M_A + M_B)^2 K_{11}[A][B]$ $[A]_{total} = [A] + K_{11}[A][B]$ $[B]_{total} = [B] + K_{11}[A][B]$
1:2 hetero-association, equivalent binding sites (e.g., Section 5.2.2)	$K_{11} = \frac{[AB]}{[A][B]}$ $K_{12} = \frac{[AB_2]}{[A][B]^2} = \left(\frac{K_{11}}{2}\right)^2$	$R/K^* = M_A^2[A] + M_B^2[B] + (M_A + M_B)^2 K_{11}[A][B]$ $+ \frac{1}{4}(M_A + 2M_B)^2 K_{11}^2[A][B]^2$ $[A]_{total} = [A] + K_{11}[A][B] + \frac{1}{4}K_{11}^2[A][B]^2$ $[B]_{total} = [B] + K_{11}[A][B] + \frac{1}{2}K_{11}^2[A][B]^2$
Isodesmic self-association to n-mers (e.g., Section 5.1.2; see Kameyama & Minton 2006)	$K_{20} = \frac{[A_2]}{[A]^2}$ $K_{n0} = \frac{[A_n]}{[A]^n} = K_{20}^{n-1}$	$R/K^* = M_A^2 \sum_{i=1}^{n} i^2 K_{20}^{i-1}[A]^i$ $[A]_{total} = \sum_{i=1}^{n} i K_{20}^{i-1}[A]^i$

Table 1. Common equilibrium association models that can be quantified by CG-MALS.

4.4.2 Kinetics models: Fitting and interpretation

Reaction kinetics for reversible and irreversible associations can be observed and quantified by light scattering to provide a direct measure of association, dissociation, or aggregation via the evolution of M_w. Quantifying characteristic rate constants from CG-MALS data requires knowledge of the final stoichiometry and, in the case of reversible associations, the appropriate equilibrium association constants. For example, LS data for covalent inhibition of an enzyme by an inhibitor may be fit at varying inhibitor concentrations to yield a second-order rate constant for the interaction, k_a. In the case of irreversible dissociation, the apparent first-order kinetics can be described by an exponential function, and the apparent dissociation rate constant, k, can be related to applicable biomolecular constants:

$$(R/K^*) \propto [A_2] = [A_2]_0\, e^{-kt} \tag{14}$$

More complex analyses, such as the association of two proteins into an equilibrium complex, involve solving the rate equations that govern the system of interest. The equilibrium association constant K_A and final stoichiometry must be determined in addition to the time-dependent change in light scattering. For the simplest heteroassociation $A + B \leftrightarrow AB$, Eq. (15) relates the CG-MALS data to the second order association rate constant $k_a = K_A \cdot k_d$:

$$\frac{1}{K*}\frac{dR}{dt} = 2M_A M_B \frac{d[AB]}{dt} ; \quad \frac{d[AB]}{dt} = k_a \left\{ \left([A]_{total} - [AB]\right)\left([B]_{total} - [AB]\right) - \frac{[AB]}{K_A} \right\} \quad (15)$$

5. CG-MALS examples

5.1 Self-association

5.1.1 Dimerization of chymotrypsin

Dimerization has been observed by CG-SLS for the enzyme α-chymotrypsin with pH-dependent affinity (Kameyama & Minton, 2006; Fernández & Minton, 2009). Figure 8 presents dependence of the reaction on ionic strength (Hanlon & Some, 2007), closely matching results obtained via sedimentation equilibrium (Aune et al., 1971).

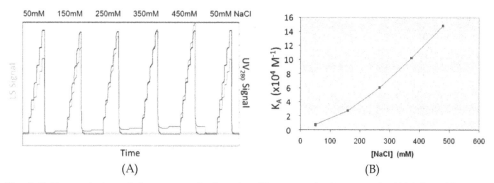

Fig. 8. Self-association of chymotrypsin forming dimers vs. ionic strength. (A) LS and UV$_{280}$ concentration data over a series of concentration gradients (B) K_A vs. [NaCl].

5.1.2 Isodesmic self-association

Some proteins tend to self assemble into chains, fibrils, or other large oligomers, such as amyloid-β plaques in Alzheimer's disease and α-synuclein aggregates in the Lewy bodies of Parkinson's disease. A model of isodesmic self-association, i.e., the assumption that each protein monomer binds to the growing chain with equal affinity, can often be used to describe such an interaction, especially in the early nucleation phase of the assembly.

Insulin changes its self-association state as a function of pH and the presence of zinc ions (Attri et al., 2010a, 2010b, and references therein). At physiological conditions in the presence of Zn^{2+}, insulin exists as a hexamer that further associates isodesmically to higher order oligomers — dimers of hexamers (12-mers), trimers of hexamers (18-mers), etc. (Attri et al., 2010b). This interaction was studied using both static and dynamic light scattering. Based on the reported equilibrium and diffusion constants, M_w, D_t, and the molar composition of insulin oligomers could be reproduced (Figure 9).

In contrast, in the absence of Zn^{2+}, insulin monomers exist in isodesmic equilibrium with dimers, trimers, and higher order complexes with pH-dependent affinity (Figure 10). Rather than constraining the maximum oligomerization state as in Table 1, both studies considered the possibility of infinitely large oligomers.

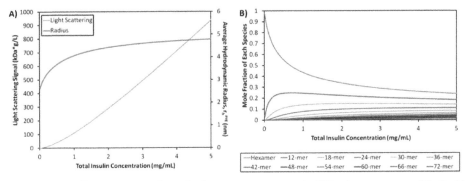

Fig. 9. Infinite self-association of insulin hexamers at neutral pH in the presence of Zn^{2+}. A) LS signal and r_h^{avg} vs. protein concentration, calculated per K_A and D_t in Attri et al., 2010b. B) Calculated molar distribution of species.

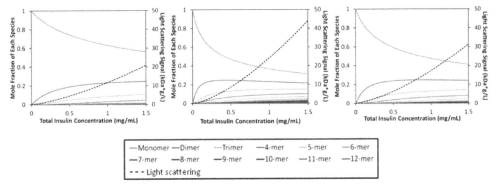

Fig. 10. Molar distribution of insulin self-association products and light scattering signal in the absence of Zn^{2+} at pH 3 (left), 7.2 (middle), and 8 (right), calculated per Attri et al., 2010a.

5.2 Hetero-association

5.2.1 Reversible enzyme-inhibitor binding with 1:1 stoichiometry

Following Kameyama & Minton 2006, we characterized a standard 1:1 reversible association between α-chymotrypsin (CT) and bovine pancreatic trypsin inhibitor (BPTI). A CG-MALS experiment consisting of self-association gradients for each binding partner CT and BPTI and a crossover hetero-association gradient was performed as per Figure 2. The self-association gradients yield the molecular weight for each monomer and confirm the lack of self-association for CT and BPTI at neutral pH. Fitting the LS data in Figure 11A as a function of composition (Figure 7B) results in an equilibrium dissociation $K_D = 119$ nM ($K_A = 8.5 \times 10^6$ M^{-1}), consistent with measurements by other techniques (referenced in Kameyama & Minton, 2006). The LS contribution from each species is then transformed to a concentration, giving the species distribution shown in Figure 11B. As expected for a 1:1 interaction, the plateau with the highest amount of complex formation occurs at a molar ratio of CT:BPTI ~1:1 (~11 µM each CT and BPTI). Since the experiment was performed at concentrations >10x K_D, nearly all available free monomer is consumed in the (CT)(BPTI) complex. This is evident in Figure 11B where the mole fraction of CT is ~0 for all compositions [CT]<[BPTI], and the mole fraction of BPTI is ~0 for all compositions [BPTI]<[CT].

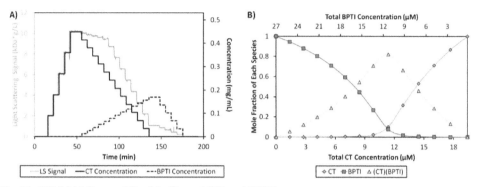

Fig. 11. CG-MALS quantifies binding of CT and BPTI.

Under acidic conditions, the affinity of CT for BPTI decreases, and CT can form reversible dimers, as in Section 5.1.1. At pH 4.4, the K_D for the association of CT and BPTI is of the same order as for the dimerization of $CT - K_D = 10$ μM and 50 μM respectively (Kameyama & Minton, 2006). Based on these results, we can simulate the expected LS signals for simultaneous self and hetero-associations (Figure 12). Discrimination between 1:1 binding only, and self + heteroassociation, is readily evident where [CT]>[BPTI] (Figure 12A). Despite the additional self-association, the fraction of hetero-association product still peaks at a molar ratio of CT:BPTI ~1:1 (Figure 12B).

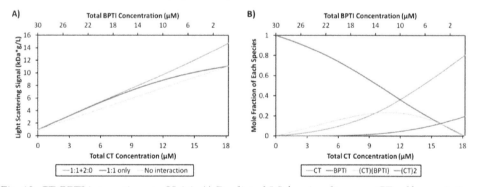

Fig. 12. CT-BPTI interaction at pH 4.4. A) Predicted LS for simultaneous CT self-association and CT-BPTI hetero-association (blue) compared to CT-BPTI hetero-association alone with $K_D = 10$ μM (red) or no interaction (green). B) Molar distribution of species for simultaneous self- and hetero-association model, based on Kameyama & Minton, 2006.

5.2.2 Antibody-antigen binding with 1:2 stoichiometry, two equivalent binding sites

The power of CG-MALS lies in its ability to identify multiple stoichiometries in solution. For example, a single multivalent receptor A may bind multiple protein ligands B, leading to the simultaneous presence of AB, AB_2, AB_3, etc. The increasing prevalence of therapeutic antibodies brings this type of multivalent binding to the forefront of biotechnology. Moreover, CG-MALS is able to characterize this type of interaction with affinities as low as $K_D\sim0.1$ nM, typical of antibody-antigen interactions. Our antibody-antigen binding data (Figure 13) indicates the presence of four species in solution: free antibody (Ab), free antigen (Ag), the 1:1 complex (Ab)(Ag), and the 1:2 complex (Ab)(Ag)$_2$. The CG-MALS K_D value of 10 nM agrees well with the literature value.

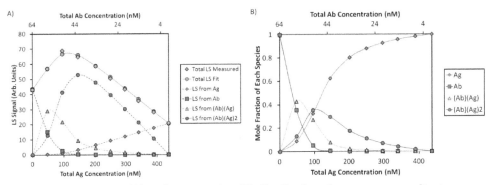

Fig. 13. Light scattering (A) and composition (B) distributions for crossover gradient between an antibody and monovalent antigen, $K_D\sim10$ nM.

Conversely, Some et al. (2008a) found that CG-MALS data for a dimeric Fcγ receptor (FcγR) binding to the Fc of a recombinant human Ab (rhumAb), shown in Figure 14, is only fit well by a model assuming two equivalent binding sites on each FcγR dimer (B) for rhumAb (A), producing equilibrium between monomers (A and B), AB, and A$_2$B (Figure 14C). CG-MALS data do not support other binding models, including 1 A : 1 B association alone (Figure 14A) and 1 A : 2 B with equivalent binding sites (Figure 14B). The calculated single-site affinity of 50 nM agrees closely with surface plasmon resonance (SPR) analysis.

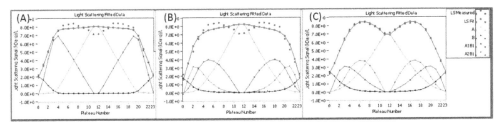

Fig. 14. Best fits (red lines) of measured CG-MALS data (blue circles) to different association models, IgG (rhumAb) : dimeric receptor (FcγR). Stoichiometry: (A) – 1:1; (B) – 1:2; (C) – 2:1. Only the {2 mAb per receptor} model fits the data.

5.2.3 Association of multivalent protein complexes

Combinations of multivalent binding partners can lead to the formation of metacomplexes in solution that are not identified by other techniques. As a homo-tetramer, streptavidin (SA) is composed of four identical binding sites capable of binding either of two Fab domains of an anti-streptavidin IgG. As we have observed, the combination of multivalent proteins enables higher-order stoichiometries to present themselves in solution, including multiple IgG molecules binding a single SA molecule and self-assemblies of IgG-SA complexes (Figure 15). Indeed, the LS signal measured for such a system by CG-MALS is nearly twice the value expected for a simple 1:2 association (Figure 15). Careful analysis of the data indicates that the solution is best described as 1:1 (IgG)(SA) complexes that self-associate (Figure 16). The infinite self-association (ISA) model employed here assumes that each base unit — (IgG)(SA) complex — assembles with other base units with the same affinity; however, this may differ from the binding-site affinity (K_D) for a single IgG-SA interaction. The binding-site K_D for one SA molecule binding one IgG was determined as 22 nM, while these 1:1 base units assemble with an average affinity K_D = 50 nM.

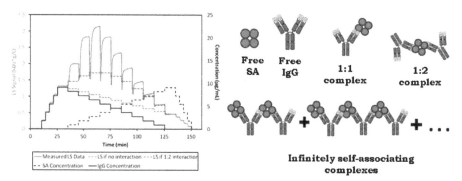

Fig. 15. Light scattering and concentration data for association of SA and anti-SA IgG. Theoretical LS plateaus are indicated for the case of no IgG-SA interaction and a 1:2 equivalent binding site model (Section 5.2.2). Additional stoichiometries that contribute to the measured LS signal, including infinitely self-associating 1:1 complexes, are shown.

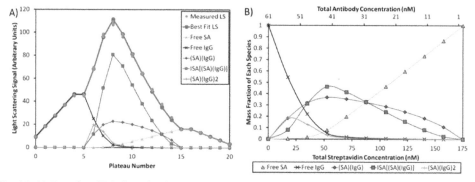

Fig. 16. A) Best fit of LS data for SA + anti-SA IgG includes infinite self assembly (ISA) of 1:1 metacomplexes B) Concentration distribution for hetero-association plateaus (#5-15).

5.3 Dissociation kinetics induced by a small molecule inhibitor

Although other techniques, such as SPR and FRET-based methods, are capable of quantifying association and dissociation kinetics, many require modification of the protein of interest, i.e., immobilization in the case of SPR and labeling with fluorescent tags for

FRET. In contrast, CG-MALS enables real-time observation of reaction kinetics in solution without protein modification. For example, chymotrypsin self-association at low pH is inhibited by the small molecule 4-(2-aminoethyl) benzenesulfonyl fluoride (AEBSF). When introduced to a chymotrypsin solution, AEBSF covalently binds the monomer active site and prevents dimerization. Varying concentrations of AEBSF were mixed with a constant stock solution of chymotrypsin, and the resulting dissociation kinetics quantified with a model of an irreversible dissociation (Some & Hanlon, 2010). For each composition, the solution was allowed to react for >1 hr while observing the decrease in weight-average molar mass of the solution. The characteristic reaction time τ ($1/k$ in Eq. (14)) varies inversely with the AEBSF concentration, consistent with the rate models defined for the system (Figure 17), indicating a rate constant of 0.064 M^{-1}s^{-1}.

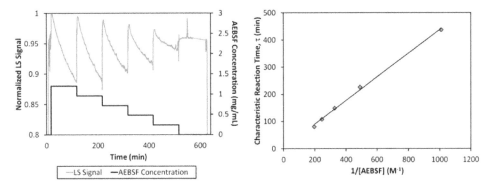

Fig. 17. Decrease in LS signal (left) and change in characteristic reaction time (right) corresponding to irreversible dissociation of chymotrypsin dimers in the presence of AEBSF.

5.4 Nonspecific interactions of non-self-associating proteins

5.4.1 Nonspecific self-interactions

As discussed in Section 2.1.4, all macromolecules at high concentrations exhibit some degree of nonspecific interactions, quantified by the second virial coefficient, A_2. This property is of particular interest in the development of pharmaceutical formulations where A_2 is one metric for the stability of a formulation and the propensity of biomolecular therapeutics to aggregate in solution. Formulations that may appear stable at moderate concentrations (~10 mg/mL or less) may indeed form self-association products at relevant formulation concentrations of 100 mg/mL or more (see Section 5.5). For a well-formulated protein, however, repulsive interactions should dominate for all concentrations of interest. BSA, for example, exhibits nonspecific repulsion even at 100 mg/mL in PBS, as shown in Fig. 18. Long-range interactions are well-screened in this buffer, resulting in an A_2=1.0x10^{-4} mol*mL/g^2, consistent with a hard-sphere of radius 3.5 nm and M$_w$ = 67 kDa.

5.4.2 Nonspecific attraction quantified by the cross-virial coefficient

Carrier proteins, drug delivery vehicles, and other polymers attract their biomolecular targets via nonspecific interactions (e.g., Dong et al., 2011) which cannot be described by an equilibrium association constant. A virial expansion may be employed to quantify nonspecific attraction or repulsion between molecules of the same species or different species. In the example below, the net negative charge of BSA, in PBS with 50 mM NaCl at pH 7, yields repulsion between BSA molecules. Lysozyme exhibits a slight positive charge with a net self-attraction as evidenced by the negative A_2. The charge-mediated attraction between BSA and lysozyme molecules is evident in Figure 19 as the increase in LS when BSA and lysozyme are mixed together. The data are best fit by a model of *nonspecific* attraction, quantified by the cross-virial coefficient A_{11}. The results can be normalized to a unitless value as per Sahin et al., 2010: $a_2 = \left(A_2^{meas} - A_2^{exc}\right)\Big/A_2^{exc}$.

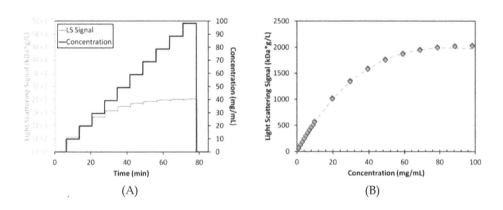

Fig. 18. BSA behaves as an effective hard sphere with $A_2 = 1.0 \times 10^{-4}$ mol*mL/g^2 for all concentrations studied. (A) CG-MALS data (B) fit to effective hard sphere model.

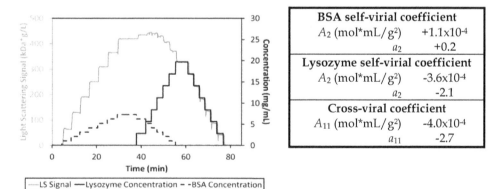

BSA self-virial coefficient	
A_2 (mol*mL/g^2)	+1.1x10^{-4}
a_2	+0.2
Lysozyme self-virial coefficient	
A_2 (mol*mL/g^2)	-3.6x10^{-4}
a_2	-2.1
Cross-virial coefficient	
A_{11} (mol*mL/g^2)	-4.0x10^{-4}
a_{11}	-2.7

Fig. 19. Determination of self-and cross-virial coefficients for nonspecific interactions in BSA-lysozyme solution. Normalized virial coefficients are also presented.

5.5 Interactions of monoclonal antibodies formulated at high concentration

Recently, CG-MALS was applied to investigate interactions between IgG1 monoclonal antibodies at concentrations up to ~300 mg/mL (Scherer et al., 2010). Although the two mAbs studied here were identical except for the CDR sequence, their self-association properties were remarkably different. MAb2 forms dimers with $K_A \leq$ ~10^3 M^{-1} ($K_D \geq$ ~1 mM), whereas mAb1 associates into dimers with K_A ~10^3-10^4 M^{-1} (K_D ~0.1-1 mM) and appears to further associate into higher order oligomers of stoichiometry 4-6. The dependence of association properties on ionic strength also differs dramatically between mAb1 and mAb2: while the affinity of the mAb2 homodimer increases with [NaCl], that of mAb1 homodimers is essentially constant. Most significantly, the higher oligomer order of mAb1 decreases from 6 to 4 as [NaCl] increased from 40 to 600 mM.

Based on these results, we reproduce in Figure 20 the relative LS signal for mAbs1 and 2 and the fraction of each oligomer present in solution. Each calculation includes the appropriate correction for non-specific repulsion using v_{eff} = 1.8 cm^3/g (r_{eff} = 4.6 nm) for mAb1 and v_{eff} = 1.4 cm^3/g (r_{eff} = 4.3 nm) for mAb2 (Scherer et al., 2010). Although the antibody molecules continue to self-associate into higher molecular weight species, the LS signal is not monotonically increasing, as would be expected from ideal scattering (Eq. (1)); instead, the LS intensity reaches a maximum at ~100 mg/mL (Figure 20A). Only by accounting for both nonspecific repulsion and specific oligomerization can the light scattering data be fully described for these high-concentration solutions.

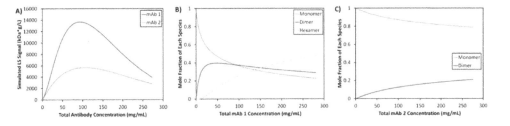

Fig. 20. A) LS signals for mAbs 1 and 2 in buffer containing 75 mM NaCl, calculated to represent results of Scherer et al., 2010. B) and C) Corresponding distribution of oligomers.

6. Conclusion

The power of light scattering, CG-MALS and CG-DLS, for investigating protein interactions lies in their great versatility. These techniques quantify a wide range of protein-protein phenomena in solution and without labeling. Both equilibrium and kinetics may be addressed directly since light scattering provides, from first principles, the molar mass and size of complexes, rather than an indirect indicator such as fluorescence. Hence light scattering is particularly well suited to analyzing higher-order complexes, multiple stoichiometries, and simultaneous self- and hetero-association. The fundamental thermodynamic nature of static light scattering provides a critical window into interactions at high concentration. The development of automation and advanced instrumentation suggests that common use of CG-MALS and CG-DLS is feasible, and hence these are important additions to the protein scientist's toolbox.

7. Acknowledgement

The authors would like to thank Allen P. Minton for many helpful discussions and collaboration in the development of automated CG-MALS; Shawn Cao (Amgen), Joey Pollastrini (Amgen), and Jihong Yang (Genentech) for contributing antibody samples; and the entire team at Wyatt Technology Corp. We are also indebted to the many early adopters of the Calypso CG-MALS system for their support and sharing samples.

8. References

Attri, A.K., Fernández, C., & Minton, A.P. (2010a). pH-dependent self-association of zinc-free insulin characterized by concentration-gradient static light scattering. *Biophysical Chemistry*. Vol. 148, No. 1-3, (May 2010), pp. 28-33, ISSN 0301-4622

Attri, A.K., Fernández, C., & Minton, A.P. (2010b). Self-association of Zn-insulin at neutral pH: investigation by concentration-gradient static and dynamic light scattering. *Biophyisical Chemistry*. Vol. 148, No. 1-3, (May 2010), pp. 23-27, ISSN 0304-4622

Attri, A.K. & Minton, A.P. (2005a). New Methods for Measuring Macromolecular Interactions in Solution via Static Light Scattering: Basic Methodology and Application to Nonassociating and Self-Associating Proteins. *Anal. Biochem.* Vol.337, No.1, (February 2005), pp. 103-110, ISSN 0003-2967

Attri, A.K. & Minton, A.P. (2005b). Composition Gradient Static Light Scattering (CG-SLS): A New Technique for Rapid Detection and Quantitative Characterization of Reversible Macromolecular Hetero-Associations in Solution. *Anal. Biochem.* Vol.337, No.1, (November 2005), pp. 103-110, ISSN 0003-2967

Aune, K.C., Goldsmith, L.C., Timasheff, S.N. (1971) Dimerization of alpha-Chymotrypsin. II. Ionic Strength and Temperature Dependence. *Biochemistry* Vol.10, No.9 (April 1971), pp. 1617-21, ISSN 0006-2960

Bajaj, H., Sharma, V.K., & Kalonia, D. (2004) Determination of Second Virial Coefficient of Proteins Using a Dual-Detector Cell for Simultaneous Measurement of Scattered Light Intensity and Concentration in SEC-HPLC. *Biophys. J.* Vol.87, No.6, (December 2004), pp. 4048-55, ISSN 0006-3495

Bajaj, H., Sharma, V.K., & Kalonia, D. (2007) A High-Throughput Method for Detection of Protein Self-Association and Second Virial Coefficient Using Size-Exclusion Chromatography Through Simultaneous Measurement of Concentration and Scattered Light Intensity. *Pharmaceutical Research* Vol.24, No.11, (November 2007), pp. 2071-83, ISSN 0724-8741

Blanco, M.A., Sahin, E., Li, Y., & Roberts, C.J. (2011) Reexamining Protein-Protein and Protein-Solvent Interactions from Kirkwood-Buff Analysis of Light Scattering in Multi-Component Solutions. *J. Chem. Phys.* Vol.134, No.22, (June 2011) pp. 225103 1-12, ISSN 0021-9606

Fernández, C. & Minton, A.P. (2008) Automated Measurement of the Static Light Scattering of Macromolecular Solutions over a Broad Range of Concentrations. *Anal. Biochem.* Vol.381, No.2, (Oct. 2008) pp. 254-7, ISSN 0003-2967

Fernández, C. & Minton, A.P. (2009) Static Light Scattering from Concentrated Protein Solutions II: Experimental Test of Theory for Protein Mixtures and Weakly Self-Associating Proteins. *Biophys. J.* Vol.96, No.5, (March 2009) pp. 1992-8, ISSN 0006-3495

Hanlon, A.D., Larkin, M.I., & Reddick, R.M. (2010) Free-Solution, Label-Free Protein-Protein Interactions Characterized by Dynamic Light Scattering. *Biophys. J.* Vol.98, No.2, (Jan 2010), p. 297-304, ISSN 0006-3495

Hanlon, A.D., and Some, D. (2007). CG-MALS for Characterization of Protein Self Association and Inhibition Kinetics. International Light Scattering Colloquium 2007, Santa Barbara, October 2007

Kameyama, K. & Minton, A.P. (2006) Rapid Quantitative Characterization of Protein Interactions by Composition Gradient Static Light Scattering. *Biophys. J.* Vol.90, No.6, (March 2006), pp. 2164-9, ISSN 0006-3495

Lehermayer, C., Mahler, H.-C., Mäder, K., & Fischer, S. (2011) Assessment of Net Charge and Protein-Protein Interactions of Different Monoclonal Antibodies. *J. Pharm. Sci.* Vol.100, No.7, (July 2011), pp. 2551-62, ISSN 0022-3549

Minton, A.P. (2007) Static Light Scattering from Concentrated Protein Solutions, I: General Theory for Protein Mixtures and Application to Self-Associating Proteins. *Biophys. J.* Vol.93, No.4, (August 2007), pp. 1321-1328, ISSN 0006-3495

Minton, A.P. and Edelhoch H. (1982) Light Scattering of Bovine Serum Albumin Solutions: Extension of the Hard Particle Model to Allow for Electrostatic Repulsion. *Biopolymers.* Vol.21, No.2, (February 1982), pp. 451-458, ISSN 0006-3525

Scherer, T.M., Liu, J., Shire, S.J. & Minton, A.P. (2010) Intermolecular Interactions of IgG1 Monoclonal Antibodies at High Concentrations Characterized by Light Scattering. *J. Phys. Chem. B.* Vol.114, No.40, (October 2010), pp. 12948-57, ISSN 1089-5647

Sahin, E., Grillo, A.O., Perkins, M.D., Roberts, C.J. Comparative Effects of pH and Ionic Strength on Protein-Protein Interactions, Unfolding, and Aggregation for AgG1 Antibodies. *J. Pharm. Sci.* Vol.99, No.12, (December 2010), pp.4830-48, ISSN 1520-6017

Some, D., Berges, A., Hitchner, E., & Yang, J. (2008a). CG-MALS Characterization of Antibody-Antigen Interactions. International Light Scattering Colloquium 2008, Santa Barbara, October 2008

Some, D. & Hanlon, A. (2010). Characterizing Protein-Protein Interactions via Static Light Scattering: Inhibition Kinetics And Dissociation. *American Biotechnology Laboratory.* Vol. 28, No. 1 (January/February 2010), pp. 9-12, ISSN 0749-3223

Some, D., Hanlon, A., & Sockolov, K. (2008b). Characterizing Protein-Protein Interactions via Static Light Scattering: Reversible Hetero-Association. *American Biotechnology Laboratory.* Vol. 26, No. 4 (March 2008), pp. 18-19, ISSN 0749-3223

Teraoka, I. (2002) *Polymer Solutions: An Introduction to Physical Properties*, John Wiley & Sons, Inc. ISBN 0-471-38929-3, New York, NY, USA

van Holde, E.; Johnson, W.C. & Ho, P. S. (1998). *Principles of Physical Biochemistry,* Prentice
    Hall, ISBN 0-13-720459-0, Upper Saddle River, NJ, USA
Young, R.J. (1981). *Introduction to Polymers,* Chapman and Hall, ISBN 0-412-22170-5, London,
    UK

8

Protein-DNA Interactions Studies with Single Tethered Molecule Techniques

Guy Nir, Moshe Lindner and Yuval Garini
*Physics Department and institute of Nanotechnology,
Bar Ilan University, Ramat Gan,
Israel*

1. Introduction

The last decade has seen a leap forward in the understanding of molecular and cellular mechanisms with the development of advanced techniques for observing, manipulating and imaging single molecules. In contrast to conventional biochemical techniques which yield information derived from population averages, single molecule techniques give access to the dynamics and properties of individual biomolecules in situ.

One of the problems in studying single molecules is the need to observe and measure the molecule for a large enough period of time and hence different approaches were developed. Among the various experimental single molecule techniques, one of the most convenient to implement is Tethered Particle Motion (TPM) (Schafer, Gelles *et al.* 1991; Nelson, Zurla *et al.* 2006; Zurla, Franzini *et al.* 2006; Jeon & Metzler 2010). In the most common procedure of TPM approach, a bead is attached to the DNA in one end, while the other end is immobilized to a glass surface (Figure 1). The Brownian motion of the nano-bead can be imaged through a microscope with an array detector such as a charged coupled detector (CCD) camera that captures the position of the bead in time and space. The bead positions are analyzed by using single particle motion (SPT) algorithms and its distribution is calculated. It is directly related to the expected conformations of the DNA when it is treated as a polymer with a given nominal length and stiffness.

The DNA contains a long chain of nucleic acids that contains all of our genetic information and a stretched human DNA molecule is about 2 meters long. This long molecule is divided to 46 chromosomes (in human cells) that are packed into a human cell that is only 10-100 μm large, which necessitates that the DNA structure will be highly regulated. Moreover, the DNA packaging has to ensure the appropriate functioning of the DNA-related processes such as transcription and replication.

All the processes that involve DNA, mainly remodeling, transcription and replication, are performed by a set of proteins that interact with the DNA in different ways. Therefore, these proteins are crucial and the understanding of their interaction patterns with the DNA is of great interest. DNA-protein interactions are therefore a subject of an ongoing research mediated by different methodologies. Better understanding of the interaction mechanisms could lead to new diagnostics methods, the discovery of new drug targets and altogether can affect mankind health.

The mechanisms described above can be studied with single molecule techniques in great details and provide information that in many cases is undetectable when using ensemble techniques. One example is the motor enzymes, such as myosin V and kinesin, that proceed in nanometric steps (Yildiz, Forkey *et al.* 2003; Yildiz, Tomishige *et al.* 2004). A detailed description of their translocation mechanism requires labeling and tracking of single enzymes in high accuracy. Observing single labeled biomolecules avoids large-population averaging and allows deciphering each step-size. It also allows distinguishing transient kinetic steps in a multistep reaction and identifying rare or short-lived conformational states.

In this book chapter we will:

1. Describe single tethered-molecule detection methods that can be used for DNA-protein studies while emphasizing Tethered Particle Motion (TPM).
2. Review few key findings on relevant proteins.
3. Compare and summarize the method presented here.
4. Provide a fascinating glance to the dazzling near-future capabilities which are based on immerging single molecule detection methods for studying DNA-protein interactions.

2. Single tethered-molecule detection methods for DNA-protein studies

Recent progress in technology, especially in the fields of photonics and nanotechnology and their combination together with the profound knowledge obtained during the last decades in molecular biology has enabled the development of single-molecule applications. Here we will describe the following well-established methods: TPM, magnetic tweezers, optical tweezers, AFM and FRET. These methods have been used to perform cutting-edge experiments as we will show later in this chapter.

2.1 Tethered particle motion

2.1.1 General description of the method

TPM is an optical method that relies on physical models and biochemical approaches to detect and observe biophysical properties, such as the dynamic variations in conformations induced by enzymes acting on biopolymers. In TPM, the biopolymer of interest, i.e. linear double-stranded DNA (dsDNA), is attached at one end to a glass surface and hence held fixed, while the other end is labeled with a marker that can be a fluorescent tag or a metal bead and is free to diffuse in a restricted volume due to the anchoring of the other end (Figure 1). The bead positions that reflect the end-to-end distance of the biopolymer are recorded at different time-intervals and are than analyzed according to physical models to retrieve biophysical properties of the biopolymer, the enzymes acting on it and the nature of their interactions. One advantage of the method relies in the fact that unlike other single molecule techniques, TPM is a force-free technique, meaning that no external force is used to alter the studied molecules natural structure, which might lead to a more reliable measurement.

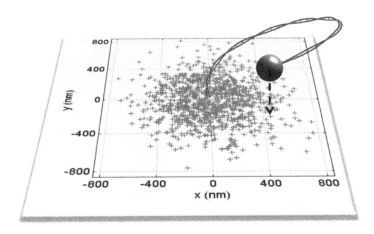

Fig. 1. TPM principles. A small bead is attached to one end of a polymer while its other end is attached to the surface. The bead performs Brownian motion in the solution constrained by the polymer. The bead positions (red crosses) are measured with an optical setup with an accuracy of few nanometers.

2.1.2 Physical model

Many biological processes such as replication, transcription and gene regulation require accessibility to the DNA. The DNA's contour in cells is regulated by DNA-bending proteins such as histones in eukaryotic cells (Shin, Santangelo *et al.* 2007) or histone-like proteins in prokaryotic cells (Rouviere-Yaniv 1987). Other proteins can bind or interact with the DNA and can modify its mechanical properties. Such interactions might alter the conformation of the DNA. TPM can detect such changes, but because they are not being measured directly, but only through the end-to-end distribution (as depicted from the bead measurement), it is necessary to use a model that describes the dependence of the DNA conformation on its basic physical parameters in space.

The DNA is known to be a rather stiff polymer that can be described to a good accuracy by the Worm-like-Chain (WLC) model. WLC is derived from the equivalent Freely Jointed Chain (FJC). The equivalent chain described by the model has the same mean square end-to-end distance $\langle R^2 \rangle$ and the same contour length, L, as the actual polymer, but it is described by N freely jointed effective bonds of length b (figure 2). This effective bond length b, is called Kuhn length. Accordingly we can write:

$$Nb = L \tag{1}$$

and its mean square end-to-end distance is

$$\langle R^2 \rangle = Nb^2 = bL . \tag{2}$$

Actually, one conventionally defines a ``persistence length", l_p, in terms of how rapidly the direction of the polymer changes as a function of the contour length. Let us define the angle θ between a vector that is tangent to a certain polymer element and a tangent vector at a distance L along the polymer. It can be shown that the expectation value of the cosine of the angle falls off exponentially with distance,

$$\langle \cos\theta \rangle = e^{-L/l_p} \tag{3}$$

where the triangular brackets denote the average over all starting polymer-element positions. For DNA, the persistence length is twice the Kuhn length.

According to the WLC theory, for "naked" (proteins-free) double stranded DNA,

$$\langle R^2 \rangle \cong 2l_p L = bL \quad \text{for} \quad L \gg l_p . \tag{4}$$

The persistence length of DNA in normal conditions is equal to 50 nm (Rubinstein & Colby 2003). If the polymer is free to assume any configuration in three dimensions, it can be shown that the probability distribution function (PDF) for its projection length along one-dimension (x or y axis) is a Gaussian:

$$P_{1D}(x)dx = \sqrt{\frac{3}{4\pi L l_p}} \cdot \exp\left(-\frac{3x^2}{4L l_p}\right) dx . \tag{5}$$

The PDF of the two-dimensional projection length of the polymer on a plane in Polar coordinates can also be calculated and it is found to be the Rayleigh distribution:

$$P_{2D}(r)dr = \frac{3}{4\pi L l_p} \cdot \exp\left(-\frac{3r^2}{4L l_p}\right) \cdot 2\pi r \cdot dr \tag{6}$$

where $r = \sqrt{x^2 + y^2}$.

Although real polymer chains are subjected to self-avoidance, meaning that due to short range repulsive forces, monomers of the chain can't cross themselves which leads to an excluded volume, it is accustomed to treat DNA molecules as ideal chains (also called phantom polymers), since the probability of crossing is non negligible only for long DNA (>40 μm) (Strick, Allemand et al.), which are not usually used in single-molecule experiments (Slutsky 2005).

2.1.3 Marker considerations

The DNA conformation changes randomly in the solution and its end-to-end distance is measured in TPM by finding the position of the attached marker. Different types of markers can be used and with respect to their detection method, one can distinguish fluorescent beads, scattering beads and beads that are detected by normal transmission or contrast enhancement methods. A fluorescent marker might be small but suffers from quenching and bleaching and is not recommended for long-time observations. Polystyrene microspheres

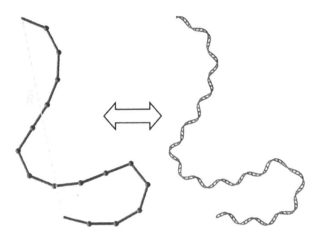

Fig. 2. Left: FJC model. Each segment is equal to b, the Kuhn length (~100 nm). $\bar{R}$ is the end-to-end vector. A real dsDNA is shown on the right and it is equivalent to the polymer in the FJC model.

are also used as markers normally with phase contrast microscopy, but it requires a rather large micron-size particle, which may lead to inaccurate analysis in TPM. If the size is too large, the measured position of the bead may be dominated by the free rotation of the marker around its tethering point. This motion is also affected by the position of the bead with respect to the surface, and therefore, the measured distribution may be dominated or severely influenced by the marker size. This will lead to errors when trying to extract the polymer properties from the bead distribution, and the bead-size effect cannot be easily compensated for. It is therefore better to work with smaller beads. The actual size that does not affect the distribution depends on the polymer contour length and persistence length. The problem was treated intensively in the literature (Segall, Nelson *et al.* 2006), and it was shown that the bead size will not affect the distribution significantly, as long as the parameter called the excursion number, is smaller than unity:

$$N_r = R / \sqrt{Ll_p / 3} < 1 . \tag{7}$$

The excursion number is defined as the ratio of the marker radius R to the radius of gyration of DNA which depends on the polymer persistence length l_p and contour length L .

Therefore, it is better to use a smaller bead size, and a more suitable solution is a small metal nano-bead. Such a bead has a significant plasmon scattering which results in an intense signal that can be easily detected by a CCD.

For DNA with a contour length of $L = 925\,nm$, a known persistence length of $l_p = 50\,nm$ for bare DNA and a gold bead with $r = 40\,nm$ gives $N_r \sim 0.32$ which meets the criterion (Segall, Nelson et al. 2006; Lindner, Nir et al. 2009).

Another advantage is the short exposure time that can be used, which still achieving a high-enough signal to noise for analyzing the bead position. If the exposure-time is too long, the bead motion broadens the image spot and the analyzed distribution of the bead position is skewed. Although this effect can be corrected (Destainville & Salomé 2006; Wong & Halvorsen 2006) it increases the error and should be avoided. We showed that with a gold nano-bead of 80 nm diameter, good results are achieved even at short exposure times as 1 ms.

2.1.4 Standard experimental set-up

Figure 3 (Nir, Lindner *et al.* 2011) shows a possible implementation of the experimental setup for TPM . It consists of a dark field (DF) microscope unit (Olympus BX-RLA2, Tokyo, Japan) with a x50 objective lens (NA=0.8) and an EM-CCD camera (Andor DU-885, Belfast, Northern Ireland) with a pixel size of 8x8 μm and a maximal pixel read-out rate of 35 MHz (Lindner, Nir *et al.* 2009).

The DF setup improves the signal to noise that is achieved in the measurement because it ensures that only the light that is scattered from the bead is collected by the objective lens, while eliminating the illumination background light.

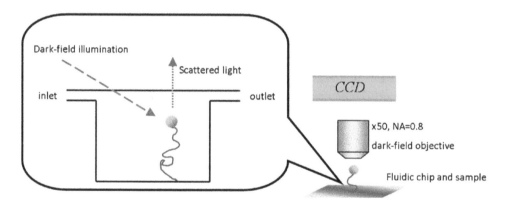

Fig. 3. The experimental setup. A dark-field microscope unit is used to track the metallic-bead position. A gold nano-bead is attached to DNA molecule and its position is tracked by the microscope and the CCD. When a protein interacts with the DNA, its biophysical properties are modified and can be tracked by the system.

2.1.5 Biochemical procedures

For constructing the dsDNA tethers it is common to use a PCR reaction to amplify desired dsDNA fragments from λ phage DNA. One end of the DNA is normally linked to digoxigenin (DIG) for a further attachment to an anti-DIG coated surface, this way the DNA is anchored to the surface. The other DNA end is biotin-linked for attaching a neutravidin conjugated nanobead. The modifications are done by using modified primers in the PCR reaction.

The tethering procedure is done in a flow cell and monitored with the microscope. First, a passivation procedure is required to reduce unspecific binding to the glass surface. Some common passivation reagents might be: Bio-Rad non-fat dry milk, Polyethylene Glycol (PEG) and Bovine Serum Albumin (BSA). After a proper incubation time (depends on the different passivation reagents) the buffer is washed and the surface is coated with anti-DIG. After one hour of incubation one should wash again and introduce the DNA to the solution. After incubation of one hour and another wash, the nano beads are introduced. After ~ 30 minutes of incubation and another wash the tethering procedure is complete (Selvin & Ha 2008; Lindner, Nir *et al.* 2009; Zimmermann, Nicolaus *et al.* 2010; Lindner, Nir *et al.* 2011).

2.1.6 Conducting experiments

First, each bead is measured a couple of times. The scattering plot is tested for circular symmetry. If it is found symmetric (figure 4), the persistence length is calculated (see section 2.1.6, data analysis). Measurement of the interaction of the bead with a protein continues only if its persistence length is ~50 ±5 nm. The next step would be to add an enzyme of interest, record the DNA's persistence length and realize if analyze the changes due to the enzyme acting on the DNA. Few beads are measured, few times each, in order to provide reliable statistic information. Each measurement consists of approximately ~2000 frames.

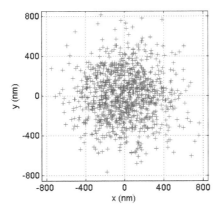

Fig. 4. XY projection. Each cross sign indicates a recording of the bead position at a time-point. Due to the symmetric polymer configurations distribution, the bead's distribution (representing the DNA end-to end vector) should be circular-symmetric centered at the DNA anchor point.

2.1.7 Data analysis

The data is analyzed with an SPT software package usually developed by the labs conducting such experiments with Matlab (The Mathworks, Natick, MA). For extraction of the DNA persistence length from the distribution, first, there should be an extraction of the bead position coordinates, $x(t), y(t)$ (2D projection) for each image (t). Then the radial distribution P(r) can be calculated and fitted to the expected distribution according to the Freely Joint Chain model which gives the Rayleigh distribution (Equation 6).

2.2 3D TPM

2.2.1 General description

3D TPM is an extended TPM method that was lately developed (Lindner, Nir *et al.* 2011). Instead of measuring only the 2D projection of the bead, it allows to measure the actual position of the bead in 3D.

The method relies on Total Internal Reflection (TIR), and employs the evanescent wave that is exponentially decreasing in the z-direction due to the TIR. Because the intensity depends on the bead height above the surface, the position of the bead in 3D can be determined.

2.2.2 Physical model

The intensity of an evanescent field decreases exponentially with the distance from the surface,

$$I = I_0 \exp(-z / d) \tag{8}$$

where I_0 is the intensity at the surface, z is the distance from surface, and d is the penetration depth. The penetration depth of the TIR illumination depends on a few fundamental parameter of the optical setup and can be tuned by changing the incident angle of the beam on the surface (Figure 5) according to:

$$d = \frac{\lambda_0}{4\pi\sqrt{n_i^2 \sin^2 \theta_i - n_t^2}} . \tag{9}$$

Here λ_0 is the wavelength of light in vacuum, n_i and n_t are the indexes of refraction of the materials above and below the surface and θ_i is the incident angle. By tuning the incident angle to be in the range of a few degrees above the critical angle, a penetration depth of 100–200 nm can be achieved, and axial distances in the range of 0–500 nm can be measured. Such a range is satisfying for measuring a 1-µm polymer with a persistence length of 50 nm; note that the end-to-end distance of such a polymer is rarely larger than 400 nm.

2.2.3 Standard experimental set-up

The experimental setup is very similar to a standard TPM setup, only with the addition of a diode laser and an equilateral prism to allow the creation of the evanescent wave (see figure 5).

2.2.4 Experimental procedures

A method that relies on the signal intensity to calculate the axial distance from the surface requires calibration in order to find d (Equation 9). We developed two calibration methods that rely on the actual system itself and do not require adding further optics to the setup. One is based on the 3D distribution of tethered beads (Volpe, Brettschneider *et al.* 2009). The distribution is measured with TIR illumination and the persistence length is calculated from the planar x and y distributions. Then, the distribution along z is fitted to the simulation results with a single parameter (the penetration depth d). The second method is based on measuring the free diffusion of the beads. The principle is similar to the fluorescence correlation spectroscopy method described by Harlepp *et al.* (Harlepp, Robert *et al.* 2004).

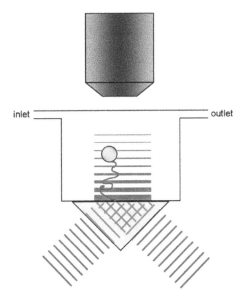

Fig. 5. 3D TPM illumination and detection. In order to achieve an evanescent wave, the light hits a prism above the critical angle. The evanescent wave decays exponentially into the sample while the bead's intensity and location is recorded. The bead position in along z is deduced according to Equation 8.

2.3 Single molecule force techniques

The following methods were thoroughly described before and hence will be briefly described here. Few key findings with these methods will be described as well.

So far we have only discussed a single molecule force-free technique. Although TPM allows us to observe a biopolymer in its natural form, it lacks the ability to manipulate the biopolymer. Many of the biological process such as DNA twisting, replication and cell migration are force driven. Another interesting fact is that biopolymers such as DNA, RNA, titin (the protein responsible for passive elasticity in the skeletal (Bustamante, Chemla *et al.* 2004)), etc posses "spring activity". If we combine these two facts, we can stretch and twist single biomolecules, hence examining their interactions with enzymes and regulating reaction coordinates as a function of load.

2.3.1 Magnetic tweezers

In magnetic tweezers a polymer of interest (such as DNA) is usually tethered to a glass surface while the other end is attached a magnetic microsphere which is pulled away from the surface with a magnet (Figure 6). The upper bound for force measurements in micromanipulation experiments is the tensile strength of a covalent bound, on the order of

1000 pN. The smallest measurable force is set by the Langevin force which is responsible for the Brownian motion of the sphere. Because of its random nature, the Langevin force is a noise density in force which is simply written as

$$f_n = \sqrt{4k_B T 6\pi\eta r} \qquad (10)$$

where η is the viscosity of the medium and r is the radius of the particle. For a ~1 µm diameter microsphere in water, $f_n \sim 0.017 \ pN / \sqrt{Hz}$. In between those two extremes lies the forces typical of the molecular scale, which is of order $k_B T / nm \sim 4pN$ (Strick, Allemand et al.). This is roughly the stall force of a single-molecular motor such as myosin (4 pN; (Finer, Simmons et al. 1994)) or RNA-polymerase (15–30 pN; (Yin, Landick et al. 1994; Wang 1998)). Applying forces in this regime on the magnetic microsphere, allows the delicate or aggressive stretching and twisting of the biopolymer, hence opening the door for manipulating single DNA-protein interactions (Manosas, Lionnet et al.; Bouchiat, Wang et al. 1999; Bustamante, Smith et al. 2000; Bustamante, Bryant et al. 2003; Neuman & Nagy 2008).

The most basic magnetic tweezers setup consists of a pair of permanent magnets, a flow cell, a magnetic bead and a CCD camera (Figure 6). For delicate manipulation, the magnet can be connected to a piezo-stage which allows bringing the magnets closer (for a stronger force) or away from the magnetic bead (for a weaker force) in nanometric steps.

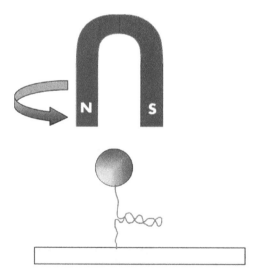

Fig. 6. Principle of magnetic tweezers. A magnetic force pulls the magnetic bead towards the magnet as a function of distance, stretching the DNA. The magnet can also be rotated, spinning the magnetic bead and twisting the DNA.

Figure 7 shows a typical plot that demonstrates how changing the force acting on the sphere allows the researchers to produce force-extension curves, meaning how much force is applied on the sphere to stretch a biopolymer to a certain extension.

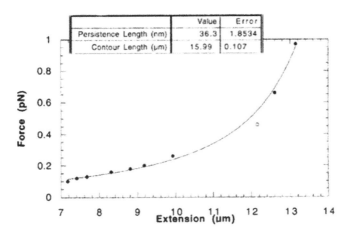

	Value	Error
Persistence Length (nm)	36.3	1.8534
Contour Length (μm)	15.99	0.107

Fig. 7. Force-extension curve. Increasing the magnetic force results in further extension of the DNA. Reprinted with permission from (Haber & Wirtz 2000). Copyright [2000]. American Institute of Physics.

The Brownian fluctuations of the tethered sphere are equivalent to the motion of a damped pendulum of length $l = <z>$ (Strick, Allemand *et al.*). Pulling the bead along z direction gives rise to a magnetic force, F. Its longitudinal, δz^2, and transverse fluctuations, δx^2, can be characterized as a spring with an effective stiffness $k_z = \partial_z F$ and $k_x = F/l$. By the equipartition theorem they satisfy:

$$\delta z^2 = \frac{k_B T}{k_z} = \frac{k_B T}{\partial_z F}$$ (11a)

and

$$\delta x^2 = \frac{k_B T}{k_x} = \frac{k_B T l}{F}.$$ (11b)

Thus by tracking the sphere fluctuations it is possible to extract the force pulling the sphere (and the biopolymer).

2.3.2 Optical tweezers

The type of experiments usually done with optical tweezers are stretching experiments, where one stretches a biomolecules of interest, for example, DNA, and follows how these manipulations alter the DNA conformation or how do DNA binding proteins respond to the load applied on the DNA. It is also used to follow the changes in extension or force as a result of a biochemical process of the DNA with proteins. The optical tweezers are implemented by creating an optical trap which is implemented by concentrating a laser beam to a diffracted-limited spot through a high Numerical Aperture (NA) objective lens. When light hits a transparent dielectric object, such as a polystyrene microsphere, it has two important optical outcomes (figure 8). The first one is reflection of the impinging light, which pushes away the microsphere (scattering force). The second one is refraction. The momentum of the light impinging on the microsphere is changed due to interaction with the microsphere. The momentum change of the light must be compensated by an equal and opposite change in momentum of the sphere, resulting in attraction of the sphere towards the center of the light spot (Williams 2002). To establish a stable trap, the force attracting the bead towards the light-spot must overcome the scattering force. The trap stiffness can be tuned by adjusting the laser intensity and focus. High NA objective lens is efficient at concentrating the light and stabilizes the trap. It is common to use high-power laser (>1 W) such as Nd:YAG and its derivatives that emit at the near-IR. The high-power reduces the spatial fluctuations of the trapped bead allowing a more stabilized trap and the near-IR wavelengths reduces the damage to the biomolecules that are in use (Neuman, Chadd *et al.* 1999). The detection of the microsphere is usually done with a quadrant photo-diode (QPD) which features high temporal resolution and leads to nanometric accuracy in detecting the deflection of the sphere (Rohrbach & AU - Stelzer 2002).

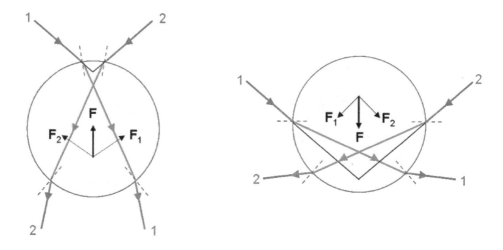

Fig. 8. Interaction of light with a transparent dielectric microsphere. Left: The microsphere is located beneath the center of the beam. When light hits the sphere it is reflected (not shown) and refracted according to Snell's law and the force acting on the object has to obey momentum conservation theory. Therefore, the net refraction force will point towards the center of the beam, pulling it up. Right: The microsphere is located above the center of the beam and the net refraction force is pointing towards the center of the beam, pulling it down.

An optical trap has to be calibrated for proper evaluating of the trap stiffness. The most common techniques treat the sphere as a linear spring, where the spring constant is determined by the sphere Brownian motion and the force obeys to Hooke's law ($f = -kx$). The sphere's position is calibrated by moving the sphere a known distance and recording the signal at that position. It can also be tuned by analyzing the frequency response of the fluctuations (Bustamante, Macosko *et al.* 2000). More profound and detailed explanations can be found in (Neuman & Nagy 2008; Selvin & Ha 2008).

As was mentioned at the beginning of this section, the most common experiments with optical tweezers are stretching experiments, which produce force-extension curves. A dsDNA molecule in solution bends and curves locally according to thermal fluctuations, which is of course an entropic-driven behavior, influenced by the DNA elasticity. According to the Worm-like Chain (WLC) model that was briefly explained in section 2.1.2, the persistence length of a dsDNA in solution containing physiological salt conditions is 50 nm (Rubinstein & Colby 2003). The WLC model is well-suited to describe the entropic behavior of dsDNA in the regime of low and intermediate forces.

According to the model, the force F required to induce an extension x of the end-to-end distance of a polymer with a contour length L and persistence length l_p is given by: (Bustamante, Smith *et al.* 2000):

$$\frac{F \cdot l_p}{k_B T} = \frac{1}{4\left(1 - x/L\right)^2} + \frac{x}{L} - \frac{1}{4} \tag{12}$$

where k_B is the Boltzmann constant and T is the temperature.

2.3.3 Atomic force microscopy with tethered molecules

The Atomic Force Microscope (AFM) (Binnig, Quate *et al.* 1986; Martin, Williams *et al.* 1987) is another force-based technique that allows the stretching of individual biomolecules (Li, Wetzel *et al.* 2006; Perez-Jimenez, Garcia-Manyes *et al.* 2006; Neuman & Nagy 2008) and measure their elastic respond with sub-nanometer accuracy and picoNewton respond. Unlike previous-discussed single-molecule force spectroscopy methods, AFM is efficient also at high forces, which opened the door for exploring the properties of bio-entities in high-energy conformational states. For example, it enabled the detection of sub-Angstrom conformational changes of a single Dextran molecule (Walther, Bruji *et al.* 2006) and plot the unfolding force-histogram of a modular polyprotein (Li, Oberhauser *et al.* 2000).

In a typical experiment, a biomolecule of interest, say a protein, is linked to a flat surface that is mounted to on a piezoelectric stage. When the protein is approached by an AFM tip which is supported by a flexible cantilever, it might adsorb to the tip. When the tip is retreated from the surface it stretches the protein. The stretch bends the cantilever, which results in deflection of the laser beam impinging on the cantilever and recorded in a photo detector. If the cantilever elastic properties are known, it is possible to extract the force acting on the protein (Fisher, Oberhauser *et al.* 1999) (see figure 9). Low-force stretching can be modeled by the WLC model and the force acting on a polymer can be extracted in the same way done with optical tweezers according to equation 12. However, it was already mentioned that the great advantage in the AFM force spectroscopy technique is the ability to apply hundreds of picoNewtons . Moreover, force-extension curves for small or single-fold proteins are difficult to interpret because of non-specific interactions that might arise between the cantilever tip and the surface.

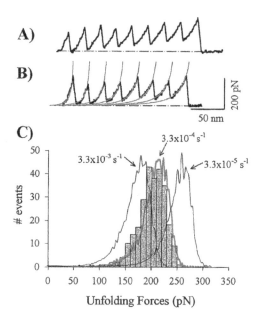

Fig. 9. Force–extension relationships for recombinant poly(I27) measured with AFM techniques. (A and B) Stretching of single I27GLG12 (A) or I27RS8 polyproteins (B) gives force–extension curves with a saw-tooth pattern having equally spaced force peaks. The saw-tooth pattern is well described by the WLC equation (continuous lines). (C) Unfolding force frequency histogram for I27RS8. The lines correspond to Monte Carlo simulations of the mean unfolding forces (n=10,000) of eight domains placed in series by using three different unfolding rate constants, k_u^0, an unfolding distance, Δx_u, of 0.25 nm, and a pulling rate of 0.6 nm/ms. Reprinted with permission from (Carrion-Vazquez, Oberhauser et al. 1999). PNAS.

To overcome this limitation and to utilize this technique advantages, modular proteins were engineered and formed "beads-on-a-string" structure (see figure 10). This structure is composed of tandem repeats of the same domain. At high forces (>100 pN), the domains are unfolded one-by-one, where each unfolding event is characterized by a saw-tooth in the force-extension curve and is explained as the elongation of the "string" due to the unfolding of a "bead" (domain) (Oberhauser, Marszalek et al. 1998). Each unfolding event can be fitted to the WLC model to recover the persistence length and the contour length of the polymer.

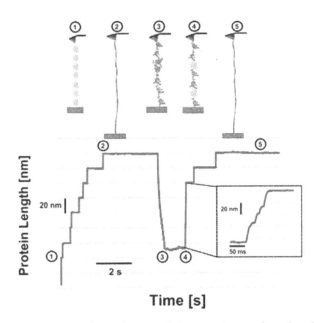

Fig. 10. 'Beads-on-a-string' stretching. At stage 1 the proteins are found at their native form (yellow). At stage 2 unfolding of the entire polyprotein occurs through pulling with the cantilever. For this protein (ubiquitin), it happens in 20 nm steps. At step 3 the force is quenched and the proteins maintain a collapsed form (gray). At step 4 refolding occur for some proteins along the chain and at step 5 the experiment is repeated by pulling again and causing another complete unfolding Reprinted with permission from (Garcia-Manyes, Dougan et al. 2009). PNAS.

2.4 Fluorescence resonance energy transfer with tethered molecules

Single-molecule FRET first introduced by Ha et al. (Ha, Enderle et al. 1996) is quite different from the other techniques discussed before. FRET aims to study the localization two entities in a biomolecule with a nanometer spatial resolution. This is done by measuring the non-radiative energy transfer from one fluorescent dye (donor) to another fluorescent dye (acceptor). The efficiency of energy transfer, E, depends on the donor-acceptor distance, R:

$$E = \frac{1}{1 + (R/R_0)^6} \tag{13}$$

Where R_0 is the distance when 50% of non-radiating energy transfer occurs ($E=0.5$) and is a function of the dyes properties (Selvin & Ha 2008). Due to the great sensitivity of this method to distance, it is applicable to a distance range of 3-8 nm. FRET is many times referred to as a spectroscopic ruler (figure 11). A biological molecule can be fluorescently-labeled in two sites, and intra-molecular dynamic motions of these sites relative to each other can be detected due to the energy transfer.

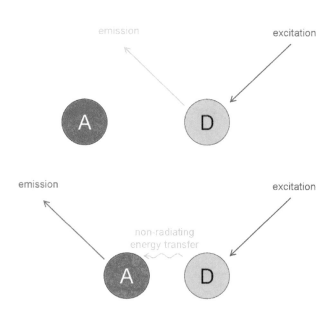

Fig. 11. FRET as a spectroscopic ruler. Up: The donor and acceptor dyes are too far for non-radiating energy transfer. Excitation of the donor results in its own emission. Down: The donor and acceptor are close enough for non-radiating energy transfer. Excitation of the donor might result in emission of the acceptor (other factors such as dipole-dipole orientation also affect the energy transfer).

Single molecule FRET (smFRET) has two major advantages over ensemble FRET. The first one derives from the fact that single-molecule experiments avoid averaging. smFRET allows the distinction between sub-populations and hence enables characterizing conformational states of biomolecules that result from dynamic and stochastic fluctuations. The second great advantage is the ability to observe in real-time dynamic events, an information that might be lost in ensemble FRET due to events being unsynchronized between different molecules.

In order to measure FRET, two detectors are needed, one for each dye. When there is a need for fast measurements and the signal is low, the preferred choice would be an Avalanche Photo-diode (APD), however choosing an Electron-Multiplying Charged Coupled Device (EMCCD) will enable visualizing hundreds of single-molecules simultaneously (Ha 2001). In smFRET the Signal-to-Noise Ratio (SNR) is relatively low (like most of the single-molecule experiments),so in order to decrease the auto fluorescence of the sample (noise) recorded by the detector, Total Internal Reflection Fluorescence microscopy (TIRF) is an appropriate solution. It ensures that only fluorescent dyes close to the surface are excited. smFRET can be applied to study surface-tethered dual-labeled DNA (McKinney, Declais *et al.* 2003) or the interactions between proteins and single tethered DNA (Myong, Rasnik *et al.* 2005; Myong, Bruno *et al.* 2007) or surface-tethered RNA (Arluison, Hohng *et al.* 2007).

3. Few key findings on relevant proteins

In this section we will show how the single molecule techniques that we described above, are employed to study a variety of biological processes such as: DNA bending by HU and IHF, DNA twisting by DNA gyrase, DNA replication by Φ29 DNA polymerase, refolding of ubiquitin and DNA unwinding by Hepatitis C virus NS3 helicase.

3.1 A DNA remodeling protein - HU

DNA-protein interactions are crucial also in bacteria cells where nucleoid-associated proteins (NAPs) together with macromolecular crowding effects play a major role in maintaining the architecture of the bacterial chromosome. NAPs ability to control the DNA structure is prominent for their role as regulators of DNA translocations (Krawiec & Riley 1990; Johnson, Johnson *et al.* 2005; Thanbichler, Wang *et al.* 2005; Luijsterburg, Noom *et al.* 2006; Stavans & Oppenheim 2006). Few of them were studied during the last decade with single molecule techniques and revealed new insights describing the dynamics of these protein-DNA/RNA interactions.

More than a few studies were performed on Integration Host Factor (IHF) protein of E. coli, which is involved in the integration of the bacteriophage λ DNA into the *E. coli* chromosome. In one of the studies, the local bending of a single 25 nm long DNA molecule caused by single IHF binding event was detected (Dixit, Singh-Zocchi *et al.* 2005). The experimental setup consisted of single linear dsDNA tethered at one end to a glass surface and to a microsphere at the other end. The tethered DNA had one consensus sequence for IHF binding. By optically monitoring the microsphere movement relative to the glass surface with evanescent microscopy, it was possible to detect conformational variations in the tether length. When IHF was introduced to the sample solution, the microsphere was pulled closer to the surface implying that the DNA bends and therefore adopts a more compact shape.

Another protein that belongs to the NAPs family is called Histone-like protein initially identified and characterized in *E.coli* strain U93 (HU). Ensemble studies revealed that the protein binds and bends DNA (Rouvière-Yaniv, Yaniv *et al.* 1979; Rouviere-Yaniv 1987; Pinson, Takahashi *et al.* 1999). Single molecule experiments refined this observation. They showed that while at relatively low HU concentrations the protein does compact the DNA, at high HU concentrations, it stretches the DNA (Sagi, Friedman *et al.* 2004; van Noort, Verbrugge *et al.* 2004; Skoko, Yoo *et al.* 2006; Xiao, Johnson *et al.* 2010). Some of these studies were performed with magnetic tweezers, (Figure 7) on a 50 nm dsDNA. Low concentrations of HU reduced the end-to-end distance in more than 50% but upon increasing the HU concentration, the persistence length increased up to ~150 nm. Similar results were also measured with a TPM setup (Figure 12). In comparison to the tweezers method, it has the advantage that force is not applied on the DNA and therefore the interaction occurs at the DNA natural form (Nir, Lindner *et al.* 2011). The bimodal effect of HU was recently explained by a model that assumes that the DNA is made of rigid segments and flexible joints (Rappaport & Rabin 2008). The model distinguishes two possible bending patterns along the polymer. If two neighboring segments are unoccupied by proteins, the bending angle θ is small, leading to the normal persistence length of DNA. When a protein occupies a segment without a neighboring protein, the spontaneous curvature increases, but when proteins occupy both neighboring segments, the spontaneous curvature is reduced again. The model therefore predicts that the DNA contains bent joints (large spontaneous curvature) and unbent joints (small spontaneous curvature) along the same DNA strand. These findings contribute to our understanding that the DNA flexibility is a more localized term when DNA-bending proteins are involved. It also raises a discussion of the contradicting role of HU. It might be, that even a low HU concentration (could be just a few nM) is efficient for chromosome condensation while higher HU concentrations might provide a degree of freedom for the interplay of a more rigid form versus chromosome condensation during different cell phases.

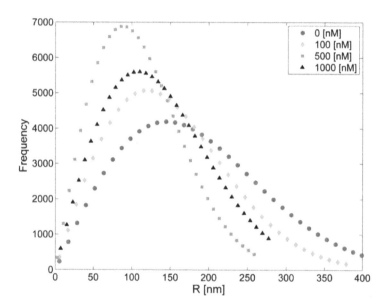

Fig. 12. Comparison of the radial distributions of the bead position for different HU concentrations as measured in a TPM experiment for the same bead. Circles represent DNA without HU proteins and the persistence length is 50 nm. Diamonds are for DNA in a solution with a concentration of 100 nM HU. The distribution is narrower compared to the HU-free DNA and the persistence length is ~39 nm. Squares represent DNA with HU concentration of 500 nM and the distribution is even narrower, with a persistence length of ~26 nm. Triangles represent DNA with HU concentration of 1000 nM and the distribution is now wider than for 500 nM with a persistence length of ~34 nm. Reprinted from Biophysical Journal **100** (2011) (Nir, Lindner *et al.* 2011) with permission from Elsevier.

So far we presented 2D TPM studies. Recently we expanded this method and made it possible to follow the DNA end-to-end distribution in 3D (Lindner, Nir *et al.* 2011). This study provided some powerful insights of the nature of the dynamic axial conformations (perpendicular to the surface) of tethered polymers. It was discovered that while the solution in the XY plane follows the normal distribution (Equation 5), the axial end-to-end distribution is different.

It can be described by a 1D random walk in half-space (Chandrasekhar 1943) and the solution is the difference between two Gaussians that centered around $\pm z_0$:

$$P(z)dz \propto \left[\exp\left(-\frac{3(z-z_0)^2}{4l_pL} \right) - \exp\left(-\frac{3(z+z_0)^2}{4l_pL} \right) \right] dz \qquad (14)$$

where z_0 is some length parameter, between the width of a DNA (2nm) and its persistence length (50nm) , and it has a negligible effect on the solution. Using the 3D TPM, we measured this distribution, which is similar to the Rayleigh distribution.

Nevertheless, it is clear from the distribution that the DNA free end is repealed from surfaces. This effect may play an important role in experiments on DNA translocation through nanopores and nuclear pores, and should affect DNA dynamics in systems where it is tethered, such as the nucleus (where it is often attached to lamins).

3.2 Studying molecular motors with single molecule techniques

As mentioned above, single-molecule force-techniques enables one to study the mechanochemical properties of specific enzymes, and more specifically also the torque and twist of the DNA. Using these methods, an upgraded magnetic tweezers setup was built for studying the twist induced by *E. Coli* DNA gyrase in DNA (Gore, Bryant *et al.* 2006). Tension was generated in a single dsDNA by pulling a magnetic microsphere attached at one end of the tethered DNA, while a 'rotor' bead (Figure 13) is attached to the center of the DNA just below an engineered single strand nick, which acts as a free swivel. The angle of the rotor bead then reflects changes in twist of the lower DNA segment, and the angular velocity of the bead is proportional to the torque in this segment. Applying tension to the DNA causes changes in linking number to partition into DNA twist, resulting in a torque on the rotor bead. An enzymatic process that changes the linking number by two will cause the rotor bead to spin around twice as the DNA returns to its equilibrium conformation. Thus, the DNA construct serves as a self-regenerating substrate for DNA gyrase. Adding *E. Coli* DNA gyrase and 1 mM ATP results in bursts of directional rotation of the rotor bead, where each burst is an even number of rotations as predicted for type II Topoisomerase, when a single catalytic cycle changes the linking number by two (Brown & Cozzarelli 1979). In order to dissect the different mechanochemical steps of the supercoiling reaction, tension was applied in the range of 0.35-1.3 pN. It was found that the supercoiling velocity doesn't vary significantly, however the processivity and initiation rate have strong dependency to template tension. As the tension increased, bursts length decreased (processivity decreased) and waiting time between bursts increased (initiation rate decreased).

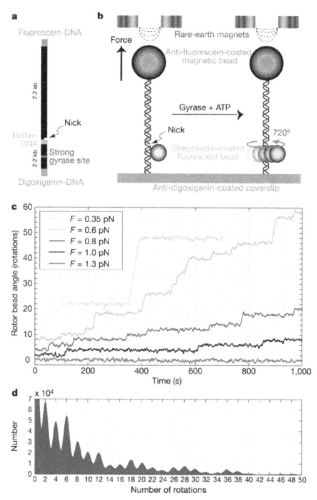

Fig. 13. Experimental design and single-molecule observations of gyrase activity. **a**, The molecular construct contains three distinct attachment sites and a site-specific nick, which acts as a swivel. A strong gyrase site was engineered into the lower DNA segment. **b**, Molecule and bead assemblies were constructed in parallel in a flow chamber and assayed by using an inverted microscope equipped with permanent magnets. Each molecule was stretched between the glass coverslip and a 1-μm magnetic bead, and a 530-nm diameter fluorescent rotor bead was attached to the central biotinylated patch. **c**, Plot of the rotor bead angle as a function of time (averaged over a 2-s window), showing bursts of activity due to diffusional encounters of individual gyrase enzymes. The activity of the enzyme is strongly dependent on tension. **d**, Histogram of the pairwise difference distribution function summed over 11 traces of 15–20 min (averaged over a 4-s window) at forces of 0.6–0.8 pN. The spacing of the peaks indicates that each catalytic cycle of the enzyme corresponds to two full rotations of the rotor bead, as expected for a type II topoisomerase such as DNA gyrase. Reprinted by permission from Macmillan Publishers Ltd: [Nature](Gore, Bryant *et al.* 2006), copyright (2006).

The measured data indicating of untwisting events caused by single DNA gyrase, allowed the researchers to build a physical model for the gyrase-DNA complex kinetics. Such a model cannot be deduced unless the single-molecule data is known.

3.3 DNA replication studies

Another study exploited the ability to apply tension on single DNA tethers using optical tweezers to investigate the conformational dynamics of the intramolecular DNA primer transfer during the processive replicative activity of the Φ29 DNA polymerase and two of its mutants (Ibarra, Chemla *et al.* 2009). Φ29 DNA polymerase has a catalytic unit as well as an exonuclease unit, allowing it to replicate DNA and fix base-pair mismatching at the same time.

The authors used optical tweezers to apply mechanical tension between two beads attached to the ends of an 8-kb dsDNA molecule with a ~400 nucleotide single-stranded gap in the middle (Figure 14A). They monitored the change in the end-to-end distance of the DNA (Δx) at constant force as the single-stranded template is replicated to dsDNA by Φ29 DNA polymerase (Figure 14B). The number of nucleotides incorporated as a function of time was obtained by dividing the observed distance change (Δx) by the expected change at a given force accompanying the conversion of one single-stranded nucleotide into its double-stranded counterpart. They also detected pause events as shown in Figure 14C.

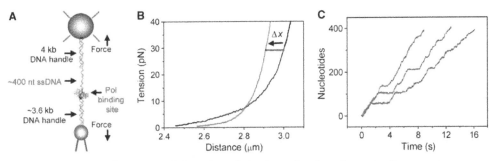

Fig. 14. Experimental set-up and detection of single-molecule polymerization events.
(A) Schematic representation of the experimental set-up (not to scale). A single DNA molecule was tethered to functionalized beads using biotin and digoxigenin moieties at the distal ends of the molecule. One bead (blue) is held in place at the end of a micropipette and the other (grey) by the optical trap. **(B)** Replication experiment (29±0.8 pN, wt polymerase) showing the force-extension curves of the initial (black) and final (red) DNA molecules. At constant force, replication shortens the distance (Δx, blue) between the beads.
(C) Representative replication traces from three independent experiments (22±0.8 pN, ed mutant). Reprinted by permission from Macmillan Publishers Ltd: [EMBO J] (Ibarra, Chemla *et al.* 2009),copyright (2009).

The authors observed an initial sharp increase in the relative pause occupancy and rationalized that it indicates that access to this intermediate from the initial pol1 (initial pol cycle) is force-sensitive and the ensuing saturation requires that the equilibrium between the intermediate and the paused state, Kp, be insensitive to the template tension. Importantly, this new intermediate must be a moving or polymerization-competent cycle (that the authors call pol2), as direct access to a non-active state in a tension-sensitive manner would lead to a continuous exponential increase in the relative pause occupancy, which was not observed.

3.4 Protein refolding studies

Understanding the dynamics of protein folding and unfolding is an ongoing effort that is presumed to benefit a lot from force spectroscopy. Single-molecule force spectroscopy techniques allow the detailed examination of the free-energy surface over which a protein diffuses in response to a mechanical perturbation (Schuler, Lipman *et al.* 2002; Rhoades, Gussakovsky *et al.* 2003). It is possible to pull a protein with an AFM tip and unfold it, a reversible process according to (Rief, Gautel *et al.* 1997; Carrion-Vazquez, Oberhauser *et al.* 1999) and upon reducing the pulling force, the unfolded protein begins to fold from a highly extended conformation that is rare or nonexistent in solution, even in the presence of denaturants. For example, at a typical force of 110 pN, mechanically unfolded ubiquitin proteins extend by >80% of their contour length (~20 nm) (Schlierf, Li *et al.* 2004). By contrast, ubiquitin proteins unfolded chemically in solution by 6 M guanidinium chloride stay compact, with a radius of gyration of only ~2.6 nm (Jacob, Krantz *et al.* 2004; Kohn, Millett *et al.* 2004).

Garcia-Manyes *et al.* studied the collapse and re-folding trajectories of ubiquitin polyproteins (Garcia-Manyes, Bruji *et al.* 2007). A chain of ubiquitin polyproteins was engineered and adsorbed to a surface while its other end was pulled by a force of 110 pN to unfold the polyproteins chain (Garcia-Manyes, Dougan *et al.* 2009). Unfolding events were observed as 20 nm steps (for each protein unfolding), followed by force-quenching to 10 pN. The time spent at the quenched state, Δt, was changed between 0.2-15 seconds before pulling again. The quenching leads to collapsed state which is mechanically unstable that is followed by folding of the protein to the native state. Therefore, changing the duration of the quenched state allows to probe the folding duration. For short durations (Δt =100-200 ms), the unfolding trajectory unravels rapidly and the 20 nm stepwise mechanism is not observed. It indicates that the proteins were not able to fold to their native state. Increasing Δt to 500 ms leads to the detection of steps when re-applying force and they predominant at Δt =3 seconds (Figure 15A). In figure 15B, a two-state process is observed, a fast initial extension that unravels the collapsed states in a stepwise manner featuring different lengths and followed by a much slower staircase of 20-nm steps, characteristic of fully refolded ubiquitin. The two states can be described by a bi-exponential fit with two rate constants, k_1 for the fast stage and k_2 for the slow state, both showing no dependency in Δt, suggesting that the protein does not gradually progress from higher to lower energy states, but populates two distinct conformational states.

The authors adopted the two-state model stating that the fast phase is a mechanically-weak state composed of a number of possible conformations with an equal distance from the transition state corresponding to the unfolding of the native state. With that knowledge, the authors tried to address the question whether these structures represent necessary folding precursors or unproductive kinetic traps in the folding energy landscape. They therefore devised a protocol to disrupt these collapsed conformations by interrupting the folding trajectories with a brief (100 ms) pulse to a higher force of 60 pN. During such a brief pulse,

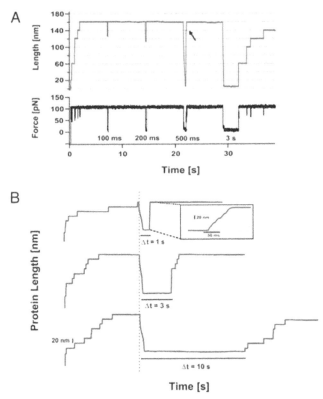

Fig. 15. Identification of a weakly stable ensemble of collapsed conformations in the folding of ubiquitin. (A) The authors repeatedly unfold and extend a ubiquitin polyprotein at 110 pN and then reduce the force to 10 pN for a varying amount of time, Δt, to trigger folding. First the polyprotein elongates in well defined steps of 20 nm, because each protein in the chain unfolds at a high force. Upon quenching the force the extended protein collapses. The state of the collapsed polypeptide was probed by raising the force back to 110 pN and measuring the kinetics of the protein elongation. (B) After full collapse the protein becomes segregated into 2 distinct ensembles: The first is identified by a fast heterogeneous elongation made of multiple sized steps (Inset); the second corresponds to well defined steps of 20 nm that identify fully folded proteins. The ratio between these 2 states of the protein depends on Δt and longer values of Δt favor the native ensemble. (Garcia-Manyes, Dougan et al. 2009). Reprinted with permission of PNAS.

native ubiquitin has a very low probability of unfolding. If the set of mechanically-unstable collapsed conformations are a prerequisite to folding, their disruption would cause a delay in the recovery of mechanical stability as compared with the unperturbed trajectories. By contrast, if the collapsed states represent unproductive traps, then unraveling them would accelerate the rate of folding. The authors showed that an average unfolding trajectory after 5 s of folding has a higher content of folded proteins than the same trajectory with the mechanical interruption. They concluded that the collapsed conformations are necessary precursors of the folded state.

3.5 Studying the unwinding of DNA by hepatitis C virus NS3 helicase

In this final example, the unwinding of DNA is demonstrated by using smFRET. In hepatitis C virus (HCV), nonstructural protein 3 (NS3) is an essential component of the viral replication complex that works with the polymerase NS5B and other protein cofactors (such as NS4A, NS5A, and NS2) to ensure effective copying of the virus. Myong *et al.* used single-molecule FRET to resolve the individual steps of DNA unwinding, catalyzed by NS3 in the absence of applied force (Myong, Bruno *et al.* 2007). Two DNA substrates were prepared. Both consisted of 18 bp and a 3'-ssDNA tail (20 nt) to create a double stranded – single stranded junction as an anchoring position to the helicase and the DNA tail was anchored to the surface. One DNA fragment had a donor and an acceptor fluorophores (cy3 and cy5 respectively) attached to the two different DNA strands at the junction through aminodeoxythymidine (figure 16A). The other DNA fragment had the dyes attached 9 bp away from the junction so that FRET signal is sensitive only to the final 9-bp unwinding

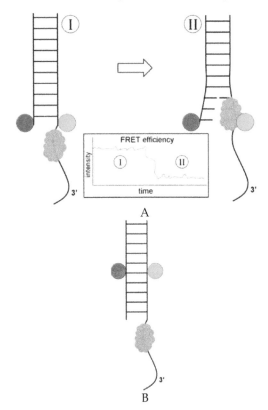

Fig. 16. DNA template for smFRET. **(A)** NS3 helicase translocates on a single dsDNA with a ssDNA tail. A green donor and a red acceptor dyes are attached at the end of the dsDNA template. At state I, the two dyes are close and the FRET efficiency (inset) is high. At state II, the initially dsDNA is partially unwound increasing the distance between the two dyes and resulting in low FRET efficiency. **(B)** In this setup the dyes are attached to the DNA in the middle of the dsDNA.

(figure 16B). Addition of ATP resulted in a decrease of the FRET efficiency as a result of strand separation due to the helicase unwinding. 6 and 3 steps were detected for the 18 bp dsDNA and 9 bp dsDNA respectively, indicating of strand separation in 3 bp steps. If hydrolysis of a single ATP results in a 3-bp step than the dwell time histogram of the steps would follow a single-exponential decay. However the observation revealed non-exponential dwell time histograms.

The authors derived a model suggesting that domains I&II of the helicase move forward one bp at a time and at the third step, the spring-loaded domain 3 moves forward in a burst motion, unzipping 3 bp as a consequence.

4. Conclusions

Single-molecule techniques are an essential tool that opened the door for high temporal and spatial studies, providing the ability to manipulate or passively observe single molecules. We presented some of the major single molecules methods and few applications of these methods for studying key biochemical processes such as DNA replication, protein folding and DNA remodeling. Table 1 summarizes the methods presented here and compares them.

	TPM	Magnetic Tweezers	Optical Tweezers	AFM	smFRET
Relies on	Probe's Brownian motion	Application of magnetic force	Momentum change of incident light	Stretching a single molecule with an AFM cantilever	Resonant Energy transfer
Probe	Nanometric gold bead	Polystyrene microsphere	Polystyrene microsphere	Cantilever	Fluorescent dyes
Range of forces (pN)	0	$10^{-3} - 10^{2}$	$10^{-1} - 10^{2}$	$10 - 10^{4}$	0
Data produced	Scattering plot (Fig.4)	Force-extension curve (Fig.7)	Force-extension curve (Fig. 14b)	Force-extension curve (Fig.16)	Energy transfer efficiency

Table 1. A comparison of single-molecule techniques reviewed here.

TPM and smFRET allows to measure without applying force and hence are good candidates to describe DNA-protein interactions in their native forms. On the other hand, force-based techniques allows to manipulate single-molecules and enable to study these interactions in extreme conditions, introducing higher energy states that are usually not observed. They reveal the mechanochemical properties of single enzymes. We showed that using TPM we were able to observe end-to-end fluctuations of single DNA molecules and measure their persistence length dynamic variations induced by HU protein. We proved that the bimodal effect of HU exists even without the use of force.

We showed how magnetic tweezers can be implemented to study the probability of DNA gyrase to achieve a productive catalytic cycle. Optical tweezers were demonstrated as a tool for discovering the proofreading activity of DNA polymerase, revealing the different steps comprising the transition from polymerase activity to exonuclease activity within the same enzyme. The use of AFM was demonstrated for catalyzing the unfolding of single proteins and then the refolding dynamics leading to the collapsed followed by the native state. Finally, smFRET was reviewed for studying helicase motor enzyme mechanisms.

All together these techniques span a broad spectrum of capabilities that can unravel different properties of single DNA-protein interactions and provide unprecedented details that are not visible by ensemble techniques. These properties includes rate constants measured directly on a single molecule, motor enzymes mechanisms, elastic and chemical properties of single molecules and high-energy states.

5. Near-future capabilities

The methodology of single molecule techniques is rapidly growing. Although the achievements that were demonstrated here, and many others, indicate on the usefulness of the existing methods, there are still important challenges that call for further improvements. High spatial resolution and high temporal resolution are necessary, especially with optical methods that can be applied to single molecules. These may be achieved with the newly developed super-resolution techniques such as stimulated depletion emission (STED), 4-pi microscopy, Photoactivated localization microscopy (PALM), stochastic optical reconstruction microscopy (STORM) and others.

Another promising direction relies on nano-optics. Due to improvements in lithography methods, it is now possible to design and fabricate nano-sized devices such as metal-based structures that use plasmonic effects. It was already shown that plasmonic nano-antennas are capable of concentrating an intense light to a sub-diffracted volume (Grigorenko, Roberts *et al.* 2008; Righini, Volpe *et al.* 2008; Huang, Maerkl *et al.* 2009; Juan, Righini *et al.* 2011) and trap dielectric particles and even *E. Coli* bacteria. These plasmonic traps do not require bulk optics (such as optical tweezers) and restrict the trapped object to a smaller volume.

Another intriguing capability is the emerging of different illumination techniques. For instance, the ability to manipulate single molecules under white light and observe fluorescent-conjugated enzymes translocating on single DNAs is fascinating and provide us with further information of the reactions nature since it enables us to follow the dynamics of both a trapped biomolecule and an enzyme translocating on it. One example is the "fleezers", fluorescent tweezers (Comstock, Ha *et al.* 2011). In this setup a dual optical trap integrated with a confocal microscope to illuminate fluorescent molecules was implemented to observe individual single fluorophore–labeled DNA oligonucleotides binding and unbinding to a complementary DNA suspended between two trapped beads.

We will finish by mentioning solid-state nanopores. Nanometer-sized holes in a thin synthetic membrane are a versatile tool for the detection and manipulation of charged biomolecules. For example, a single DNA molecule that translocates through the nanopore will have a unique signature that is attributed to its sequence. That can be done by applying an external electric field which drives a biomolecule through the nanopore, producing a characteristic transient change in the trans-pore ionic current (Heng, Ho *et al.* 2004; Garaj, Hubbard *et al.* 2010; Stefan, Alexander *et al.* 2011).

6. Acknowledgements

This work was supported in part by the Israel Science Foundation grants 985/08, 1729/08, 1793/07, and 25/07.

7. References

Arluison, V., S. Hohng, R. Roy, O. Pellegrini, P. Regnier & T. Ha (2007). Spectroscopic Observation of RNA Chaperone Activities of Hfq in Post-Transcriptional Regulation by a Small Non-Coding RNA. *Nucl. Acids Res*, 353, pp. 999-1006

Binnig, G., C. F. Quate & C. Gerber (1986). Atomic Force Microscope. *Phys. Rev. Lett*, 569, pp. 930

Bouchiat, C., M. D. Wang, J.-F. Allemand, T. Strick, S. M. Block & V. Croquette (1999). Estimating the Persistence Length of a Worm-Like Chain Molecule from Force-Extension Measurements. *Biophys. J*, 76, pp. 409-413

Brown, P. O. & N. R. Cozzarelli (1979). A Sign Inversion Mechanism for Enzymatic Supercoiling of DNA. *Science*, 2064422, pp. 1081-1083

Bustamante, C., Z. Bryant & S. B. Smith (2003). Ten Years of Tension: Single-Molecule DNA Mechanics. *Nature*, 4216921, pp. 423-427

Bustamante, C., Y. R. Chemla, N. R. Forde & D. Izhaky (2004). Mechanical Processes in Biochemistry. *Annu. Rev. Biochem*, 731, pp. 705-748

Bustamante, C., J. C. Macosko & G. J. L. Wuite (2000). Grabbing the Cat by the Tail: Manipulating Molecules One by One. *Nat. Rev. Mol. Cell Biol*, 1, pp. 131-136

Bustamante, C., S. B. Smith, J. Liphardt & D. Smith (2000). Single-Molecule Studies of DNA Mechanics. *Curr. Opin. Struct. Biol*, 103, pp. 279-285

Carrion-Vazquez, M., A. F. Oberhauser, S. B. Fowler, P. E. Marszalek, S. E. Broedel, J. Clarke & J. M. Fernandez (1999). Mechanical and Chemical Unfolding of a Single Protein: A Comparison. *Proc. Natl. Acad. Sci. U S A*, 967, pp. 3694-3699

Chandrasekhar, S. (1943). Stochastic Problems in Physics and Astronomy. *Rev. Mod. Phys*, 151, pp. 1

Comstock, M. J., T. Ha & Y. R. Chemla (2011). Ultrahigh-Resolution Optical Trap with Single-Fluorophore Sensitivity. *Nat. Meth*, 84, pp. 335-340

Destainville, N. & L. Salomé (2006). Quantification and Correction of Systematic Errors Due to Detector Time-Averaging in Single-Molecule Tracking Experiments *Biophys. J*, 902, pp. L17-L19

Dixit, S., M. Singh-Zocchi, J. Hanne & G. Zocchi (2005). Mechanics of Binding of a Single Integration-Host-Factor Protein to DNA. *Phys. Rev. Lett*, 9411, pp. 118101

Finer, J. T., R. M. Simmons & J. A. Spudich (1994). Single Myosin Molecule Mechanics: Piconewton Forces and Nanometre Steps. *Nature*, 3686467, pp. 113-119

Fisher, T. E., A. F. Oberhauser, M. Carrion-Vazquez, P. E. Marszalek & J. M. Fernandez (1999). The Study of Protein Mechanics with the Atomic Force Microscope. *Trends Biochem. Sci*, 2410, pp. 379-384

Garaj, S., W. Hubbard, A. Reina, J. Kong, D. Branton & J. A. Golovchenko (2010). Graphene as a Subnanometre Trans-Electrode Membrane. *Nature*, 4677312, pp. 190-193

Garcia-Manyes, S., Bruji, cacute, Jasna, C. L. Badilla & J. M. Fernandez (2007). Force-Clamp Spectroscopy of Single-Protein Monomers Reveals the Individual Unfolding and Folding Pathways of I27 and Ubiquitin. *Biophys. J*, 937, pp. 2436-2446

Garcia-Manyes, S., L. Dougan, C. L. Badilla, J. Bruji & J. M. Fernandez (2009). Direct Observation of an Ensemble of Stable Collapsed States in the Mechanical Folding of Ubiquitin. *Proc. Natl. Acad. Sci. U S A*,

Gore, J., Z. Bryant, M. D. Stone, M. Nollmann, N. R. Cozzarelli & C. Bustamante (2006). Mechanochemical Analysis of DNA Gyrase Using Rotor Bead Tracking. *Nature*, 4397072, pp. 100-104

Grigorenko, A. N., N. W. Roberts, M. R. Dickinson & Y. Zhang (2008). Nanometric Optical Tweezers Based on Nanostructured Substrates. *Nat. Photon*, 26, pp. 365-370

Ha, T. (2001). Single-Molecule Fluorescence Resonance Energy Transfer. *Methods*, 25, pp. 78-86

Ha, T., T. Enderle, D. F. Ogletree, D. S. Chemla, P. R. Selvin & S. Weiss (1996). Probing the Interaction between Two Single Molecules: Fluorescence Resonance Energy Transfer between a Single Donor and a Single Acceptor. *Proc. Natl. Acad. Sci. U S A*, 93, pp. 6264–6268

Haber, C. & D. Wirtz (2000). Magnetic Tweezers for DNA Micromanipulation. *Rev. Sci. Instrum.*, 7112, pp. 4561

Harlepp, S., J. Robert, N. C. Darnton & D. Chatenay (2004). Subnanometric Measurements of Evanescent Wave Penetration Depth Using Total Internal Reflection Microscopy Combined with Fluorescent Correlation Spectroscopy. *Applied Physcis Letters*, 85, pp. 3917-3919

Heng, J. B., C. Ho, T. Kim, R. Timp, A. Aksimentiev, Y. V. Grinkova, S. Sligar, K. Schulten & G. Timp (2004). Sizing DNA Using a Nanometer-Diameter Pore. *Biophys. J*, 874, pp. 2905-2911

Huang, L., S. J. Maerkl & O. J. Martin (2009). Integration of Plasmonic Trapping in a Microfluidic Environment. *Opt. Express*, 178, pp. 6018-6024

Ibarra, B., Y. R. Chemla, S. Plyasunov, S. B. Smith, J. M. Lazaro, M. Salas & C. Bustamante (2009). Proofreading Dynamics of a Processive DNA Polymerase. *EMBO J*, 2818, pp. 2794-2802

Jacob, J., B. Krantz, R. S. Dothager, P. Thiyagarajan & T. R. Sosnick (2004). Early Collapse Is Not an Obligate Step in Protein Folding. *J. Mol. Biol*, 3382, pp. 369-382

Jeon, J.-H. & R. Metzler (2010). Fractional Brownian Motion and Motion Governed by the Fractional Langevin Equation in Confined Geometries. *Phys. Rev. E*, 812, pp. 021103

Johnson, R. C., L. M. Johnson, J. W. Schmidt & J. F. Garder (2005). Major Nucleoid Proteins in the Structure and Function of the *Escherichia Coli* Chromosome. In: *The Bacterial Chromosome*. N. P. Higgins. 1: 65-132, American Society for Microbiology. Washington, DC.

Juan, M. L., M. Righini & R. Quidant (2011). Plasmon Nano-Optical Tweezers. *Nat. Photon*, 56, pp. 349-356

Kohn, J. E., I. S. Millett, *et al.* (2004). Random-Coil Behavior and the Dimensions of Chemically Unfolded Proteins. *Proc. Natl. Acad. Sci. U S A*, 10134, pp. 12491-12496

Krawiec, S. & M. Riley (1990). Organization of the Bacterial Chromosome. *Microbiol. Rev.*, 544, pp. 502-539

Li, H., A. F. Oberhauser, S. B. Fowler, J. Clarke & J. M. Fernandez (2000). Atomic Force Microscopy Reveals the Mechanical Design of a Modular Protein. *Proc. Natl. Acad. Sci. U S A*, 9712, pp. 6527-6531

Li, L., S. Wetzel, A. Pluckthun & J. M. Fernandez (2006). Stepwise Unfolding of Ankyrin Repeats in a Single Protein Revealed by Atomic Force Microscopy. *Biophys. J*, 904, pp. L30-L32

Lindner, M., G. Nir, H. R. C. Dietrich, Ian T. Young, E. Tauber, I. Bronshtein, L. Altman & Y. Garini (2009). Studies of Single Molecules in Their Natural Form. *Isr. J. Chem*, 493-4, pp. 283-291

Lindner, M., G. Nir, S. Medalion, H. R. C. Dietrich, Y. Rabin & Y. Garini (2011). Force-Free Measurements of the Conformations of DNA Molecules Tethered to a Wall. *Phys. Rev. E*, 831, pp. 011916

Luijsterburg, M. S., M. C. Noom, G. J. L. Wuite & R. T. Dame (2006). The Architectural Role of Nucleoid-Associated Proteins in the Organization of Bacterial Chromatin: A Molecular Perspective. *J. Struct. Biol*, 1562, pp. 262-272

Manosas, M., T. Lionnet, E. Praly, D. Fangyuan, J.-F. Allemand, D. Bensimon, V. Croquette & V. Rivasseau Studies of DNA-Replication at the Single Molecule Level Using Magnetic Tweezers. *Biological Physics*, 60, pp. 89-122

Martin, Y., C. C. Williams & H. K. Wickramasinghe (1987). Atomic Force Microscope - Force Mapping and Profiling on a Sub 100 Å Scale. *J. Appl. Phys*, 6110, pp. 4723-4729

McKinney, S. A., A.-C. Declais, D. M. J. Lilley & T. Ha (2003). Structural Dynamics of Individual Holliday Junctions. *Nat. Struct. Mol. Biol*, 102, pp. 93-97

Myong, S., M. M. Bruno, A. M. Pyle & T. Ha (2007). Spring-Loaded Mechanism of DNA Unwinding by Hepatitis C Virus Ns3 Helicase. *Science*, 3175837, pp. 513-516

Myong, S., I. Rasnik, C. Joo, T. M. Lohman & T. Ha (2005). Repetitive Shuttling of a Motor Protein on DNA. *Nature*, 4377063, pp. 1321-1325

Nelson, P. C., C. Zurla, D. Brogioli, J. F. Beausang, L. Finzi & D. Dunlap (2006). Tethered Particle Motion as a Diagnostic of DNA Tether Length. *J. Phys. Chem. B*, 11034, pp. 17260-17267

Neuman, K. C., E. H. Chadd, G. F. Liou, K. Bergman & S. M. Block (1999). Characterization of Photodamage to Escherichia Coli in Optical Traps. *Biophys. J*, 775, pp. 2856-2863

Neuman, K. C. & A. Nagy (2008). Single-Molecule Force Spectroscopy: Optical Tweezers, Magnetic Tweezers and Atomic Force Microscopy. *Nat. Meth*, 56, pp. 491-505

Nir, G., M. Lindner, H. R. C. Dietrich, O. Girshevitz, C. E. Vorgias & Y. Garini (2011). HU Protein Induces Incoherent DNA Persistence Length. *Biophys. J*, 100, pp. 784-90

Oberhauser, A. F., P. E. Marszalek, H. P. Erickson & J. M. Fernandez (1998). The Molecular Elasticity of the Extracellular Matrix Protein Tenascin. *Nature*, 3936681, pp. 181-185

Perez-Jimenez, R., S. Garcia-Manyes, S. R. K. Ainavarapu & J. M. Fernandez (2006). Mechanical Unfolding Pathways of the Enhanced Yellow Fluorescent Protein Revealed by Single Molecule Force Spectroscopy. *J. Biol. Chem*, 28152, pp. 40010-40014

Pinson, V., M. Takahashi & J. Rouviere-Yaniv (1999). Differential Binding of the *Escherichia Coli* HU, Homodimeric Forms and Heterodimeric Form to Linear, Gapped and Cruciform DNA. *J. Mol. Biol*, 287, pp. 485-497

Rappaport, S. M. & Y. Rabin (2008). Model of DNA Bending by Cooperative Binding of Proteins. *Phys. Rev. Lett*, 1013, pp. 038101

Rhoades, E., E. Gussakovsky & G. Haran (2003). Watching Proteins Fold One Molecule at a Time. *Proc. Natl. Acad. Sci. U S A*, 1006, pp. 3197-3202

Rief, M., M. Gautel, F. Oesterhelt, J. M. Fernandez & H. E. Gaub (1997). Reversible Unfolding of Individual Titin Immunoglobulin Domains by AFM. *Science*, 2765315, pp. 1109-1112

Righini, M., G. Volpe, C. Girard, P. Dimitri & R. Quidant (2008). Surface Plasmon Optical Tweezers: Tunable Optical Manipulation in the Femtonewton Range. *Phys. Rev. Lett*, 10018, pp. 186804

Rohrbach, A. & E. AU - Stelzer (2002). Three-Dimensional Position Detection of Optically Trapped Dielectric Particles. *J. Appl. Phys.*, 918,

Rouvière-Yaniv, J., M. Yaniv & J.-E. Germond (1979). E. Coli DNA Binding Protein HU Forms Nucleosome-Like Structure with Circular Double-Stranded DNA. *Cell*, 172, pp. 265-274

Rouviere-Yaniv, K. D. a. J. (1987). Histonelike Proteins of Bacteria. *Microbiological review*, 513, pp. 19

Rubinstein, M. & R. H. Colby (2003). Polymer Physics, Oxford University Press.

Sagi, D., N. Friedman, C. Vorgias, A. B. Oppenheim & J. Stavans (2004). Modulation of DNA Conformations through the Formation of Alternative High-Order HU-DNA Complexes. *J. Mol. Biol*, 3412, pp. 419-428

Schafer, D. A., J. Gelles, M. P. Sheetz & R. Landick (1991). Transcription by Single Molecules of RNA Polymerase Observed by Light Microscopy. *Nature*, 352, pp. 444-448

Schlierf, M., H. Li & J. M. Fernandez (2004). The Unfolding Kinetics of Ubiquitin Captured with Single-Molecule Force-Clamp Techniques. *Proc. Natl. Acad. Sci. U S A*, 10119, pp. 7299-7304

Schuler, B., E. A. Lipman & W. A. Eaton (2002). Probing the Free-Energy Surface for Protein Folding with Single-Molecule Fluorescence Spectroscopy. *Nature*, 4196908, pp. 743-747

Segall, D. E., P. C. Nelson & R. Phillips (2006). Volume-Exclusion Effects in Tethered-Particle Experiments: Bead Size Matters. *Phys. Rev. Lett*, 96, pp. 0883061-4

Selvin, P. R. & T. Ha (2008). Single-Molecule Techniques: A Laboratory Manual, Cold Spring Harbor Laboratory Press.

Shin, J.-H., T. J. Santangelo, Y. Xie, J. N. Reeve & Z. Kelman (2007). Archaeal Minichromosome Maintenance (Mcm) Helicase Can Unwind DNA Bound by Archaeal Histones and Transcription Factors. *J. Biol. Chem*, 2827, pp. 4908-4915

Skoko, D., D. Yoo, H. Bai, B. Schnurr, J. Yan, S. M. McLeod, J. F. Marko & R. C. Johnson (2006). Mechanism of Chromosome Compaction and Looping by the Escherichia Coli Nucleoid Protein Fis. *J. Mol. Biol*, 3644, pp. 777-798

Slutsky, M. (2005). Diffusion in a Half-Space: From Lord Kelvin to Path Integrals. *Am. J. Phys*, 734, pp. 308-314

Stavans, J. & A. Oppenheim (2006). DNA-Protein Interactions and Bacterial Chromosome Architecture. *Phys. Biol*, 34, pp. R1-10

Kowalczyk, S. W., Y. A. Grosberg, Y. Rabin & C. Dekker (2011). Modeling the Conductance and DNA Blockade of Solid-State Nanopores. *Nanotechnology*, 2231, pp. 315101

Strick, T., J.-F. Allemand, V. Croquette & D. Bensimon Twisting and Stretching Single DNA Molecules. *Prog. Biophys. Mol. Biol*, 741-2, pp. 115-140

Thanbichler, M., S. C. Wang & L. Shapiro (2005). The Bacterial Nucleoid: A Highly Organized and Dynamic Structure. *J. Cell. Biochem*, 963, pp. 506-521

van Noort, J., S. Verbrugge, N. Goosen, C. Dekker & R. T. Dame (2004). Dual Architectural Roles of HU: Formation of Flexible Hinges and Rigid Filaments. *Proc. Natl. Acad. Sci. U S A*, 10118, pp. 6969-6974

Volpe, G., T. Brettschneider, L. Helden & C. Bechinger (2009). Novel Perspectives for the Application of Total Internal Reflection Microscopy. *Opt. Express*, 1726, pp. 23975-23985

Walther, K. A., J. Bruji, H. Li & J. M. Fernandez (2006). Sub-Angstrom Conformational Changes of a Single Molecule Captured by Afm Variance Analysis. *Biophys. J*, 9010, pp. 3806-3812

Wang, J. C. (1998). Moving One DNA Double Helix through Another by a Type Ii DNA Topoisomerase: The Story of a Simple Molecular Machine. *Q. Rev. Biophys*, 3102, pp. 107-144

Williams, M. C. (2002). Optical Tweezers: Measuring Piconewton Forces, Biophysical Society.

Wong, W. P. & K. Halvorsen (2006). The Effect of Integration Time on Fluctuation Measurements: Calibrating an Optical Trap in the Presence of Motion Blur. *Opt. Express*, 1425, pp. 12517-12531

Xiao, B., R. C. Johnson & J. F. Marko (2010). Modulation of Hu-DNA Interactions by Salt Concentration and Applied Force. *Nucl. Acids Res*, 3818, pp. 6176-6185

Yildiz, A., J. N. Forkey, S. A. McKinney, T. Ha, Y. E. Goldman & P. R. Selvin (2003). Myosin V Walks Hand-over-Hand: Single Fluorophore Imaging with 1.5-Nm Localization. *Science*, 3005628, pp. 2061-2065

Yildiz, A., M. Tomishige, R. D. Vale & P. R. Selvin (2004). Kinesin Walks Hand-over-Hand. *Science*, 3035658, pp. 676-678

Yin, H., R. Landick & J. Gelles (1994). Tethered Particle Motion Method for Studying Transcript Elongation by a Single RNA Polymerase Molecule. *Biophys. J*, 67, pp. 2468-2478

Zimmermann, J. L., T. Nicolaus, G. Neuert & K. Blank (2010). Thiol-Based, Site-Specific and Covalent Immobilization of Biomolecules for Single-Molecule Experiments. *Nat. Protocols*, 56, pp. 975-985

Zurla, C., A. Franzini, G. Galli, D. D. Dunlap, D. E. A. Lewis, S. Adhya & L. Finzi (2006). Novel Tethered Particle Motion Analysis of Ci Protein-Mediated DNA Looping in the Regulation of Bacteriophage Lambda. *J. Phys.: Condens. Matter*, 18, pp. S225-S234

9

Site-Directed Spin Labeling and Electron Paramagnetic Resonance (EPR) Spectroscopy: A Versatile Tool to Study Protein-Protein Interactions

Johann P. Klare

Physics Department, University of Osnabrück, Osnabrück
Germany

1. Introduction

The function of a living cell, independent of we are talking about a prokaryotic single-cellular organism or a cell in the context of an complex organism like a human, depends on intricate and balanced interaction between its components. Proteins are playing a central role in this complex cellular interaction network: Proteins interact with nucleic acids, with membranes of all cellular compartments, and, what will be in the focus of this article, with other proteins. Proteins interact to form functional units, to transmit signals for example perceived at the surface of the cell to cytoplasmic or nuclear components, or to target them to specific locations. Thus, the study of protein-protein interactions on the molecular level provides insights into the basic functional concepts of living cells and emerged as a wide field of intense research, steadily developing with the introduction of new and refined biochemical and biophysical methods.

Nowadays there is a vast of methods available to study the interaction between proteins. On the biochemical level mutational studies, crosslinking experiments and chromatographic techniques provide means to identify and characterize the interfaces on the protein surface where interaction takes place. Biophysical methods include calorimetric techniques, fluorescence spectroscopy and microscopy, and "structural techniques" like X-ray crystallography, (cryo-) electron microscopy, NMR spectroscopy, FRET spectroscopy, and EPR spectroscopy on spin labeled proteins.

Site-directed spin labeling (SDSL) (Altenbach et al., 1989a, 1990) in combination with electron paramagnetic resonance (EPR) spectroscopy has emerged as a powerful tool to investigate the structural and the dynamical aspects of biomolecules, under conditions close to physiological i.e. functional state of the system under exploration. The technique is applicable to soluble molecules and membrane bound proteins either solubilised in detergent or embedded in a lipid bilayer. Therein, the size and the complexity of the system under investigation is almost arbitrary (reviewed in Bordignon & Steinhoff, 2007; Hubbell et al., 1996; Hubbell et al., 1998; Klare & Steinhoff, 2009; Klug & Feix, 2007). Especially with respect to protein-protein interactions SDSL EPR can provide a vast amount of information about almost all aspects of this interaction. Spin labeling approaches can provide detailed information about the binding interface not only on the structural level but also give insights into kinetic and thermodynamic aspects of the interaction. EPR also allows determination of distances between pairs of spin labels in the range from ~ 10-80 Å with accuracies down to less than 1 Å, thereby covering a range of sizes including also large multi-domain proteins and protein complexes.

This chapter will give an introduction into the technique of SDSL EPR spectroscopy exemplified with data from studies on the photoreceptor/transducer-complex NpSRII/NpHtrII, followed by a number of recent examples from the literature where protein-protein interactions have been studied using this technique.

2. Site-Directed Spin Labeling (SDSL)

2.1 Spin labeling of cysteines

For the modification proteins with spin labels, three different approaches have been established. The most commonly used method utilizes the reactivity of the sulfhydryl group of cysteine residues being engineered into the protein applying site-directed mutagenesis. This approach usually requires that the protein of interest possesses only cysteine residues at the desired sites, and that additional cysteine residues present can be replaced by serines or alanines without impairment of protein functionality. Among the various spin labels available the (1-oxyl-2,2,5,5-tetramethylpyrroline-3-methyl) methanethiosulfonate spin label (MTSSL) (Berliner et al., 1982) is most often used due to its sulfhydryl specificity, and its small molecular volume comparable to that of a tryptophane side chain (Fig. 1A,B). The spin label is bound to the protein by formation of a disulfide bond with the cysteine, and the resulting spin label side chain is commonly abbreviated R1. The linker between the nitroxide ring and the protein backbone renders the R1 side chain flexible (Fig. 1B), thus minimizing disturbances of the native fold and the function of the protein it is attached to. In addition, the unique dynamic properties of this spin-label side chain provide detailed structural information from the shape of its room temperature EPR spectrum. Besides MTSSL, a variety of different nitroxide radical compounds are commercially available, for example the 1-oxyl-2,2,5,5-tetramethyl, 2,5-dihydro-1H-pyrrol-3-carboxylic acid (2-methanethiosulphonyl-ethyl) amide (MTS-4-oxyl) spin label (Fig. 1C), comprising different linkers and/or nitroxide moieties. Also pH sensitive spin probes have been used to label the thiol group, e.g. of a synthetic peptide fragment of the laminin B1 chain (Smirnov et al., 2004).

The widely used methanethiosulfonate spin labels suffer from a significant drawback which is the sensitivity of the disulfide bond towards reducing agents like DTT, leading to immediate release of spin label side chains. If reducing conditions are required for sample preparation and/or stability, acetamide or maleimide-functionalized spin label compounds (Steinhoff et al., 1991; Griffith and McConnell, 1966) (Fig. 1D) can be used alternatively. In this case, the spin label is bound via a CS bond, which is not affected by reducing conditions.

Isotopically labeled nitroxide compounds where ^{14}N is exchanged by ^{15}N are important for specialized applications. The corresponding EPR spectra are characterized by a two line spectrum instead of a three line spectrum of the ^{14}N, and the lines of a ^{15}N spectrum are well

Site-Directed Spin Labeling and Electron Paramagnetic Resonance (EPR) Spectroscopy: A Versatile Tool to Study Protein-Protein Interactions

179

separated from the ^{14}N lines so that both labels can be used simultaneously in a single experiment (Steinhoff et al., 1991).

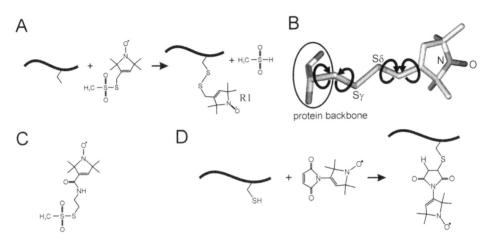

Fig. 1. Spin labeling of cysteines (A) Reaction of the methanethiosulfonate spin label (MTSSL) with a cysteine side chain, generating the spin label side chain R1. (B) Flexible bonds within the R1 side chain are indicated. (C) Chemical structure of the MTS-4-oxyl spin label. (D) Reaction of a maleimide spin label N-(1-oxyl-2,2,6,6-tetramethyl-4-piperidinyl)maleimide with a cysteine side chain.

2.2 Spin labeling by peptide synthesis

A large variety of spin label building blocks for Boc- or Fmoc-based step-by-step peptide synthesis either on a solid support (SPPS) (Merrifield, 1963) or in solution have been synthesized (Barbosa et al., 1999; Elsässer et al., 2005). Being the most popular one, the paramagnetic α-amino acid TOAC (4-amino-1-oxyl-2,2,6,6,-tetramethyl-piperidine-4-carboxylic acid) (Rassat and Rey, 1967) is characterized by only one degree of freedom, the conformation of the six-membered ring (Fig. 2) The nitroxide is rigidly coupled to the peptide backbone, thereby providing the possibility to obtain direct information about the orientation of secondary structure elements, and has for example been used to study the secondary structure of small peptides in liquid solution (Anderson et al., 1999; Hanson et al., 1996; Marsh et al., 2007), and has also been successfully incorporated into the α-melanocyte stimulating hormone without loss of function (Barbosa et al., 1999).

The chemical synthesis of proteins with incorporated unnatural spin labeled amino acids relies on the ability to produce the constituent peptides, typically by SPPS. Although synthesis of polypeptides consisting of more than 160 amino acids (Becker et al., 2003) has become possible through improvements in peptide chemistry, aiming at the incorporation of spin labels into large proteins, esp. membrane proteins, SPPS has to be combined with recombinant techniques. The expressed protein ligation (EPL), also named intein mediated protein ligation (IPL) technique, can be used to semisynthesize proteins from recombinant and synthetic fragments, thereby extending the size and complexity of the protein targets.

The underlying chemical ligation of two polypeptide fragments requires an N-terminal cysteine on one and a C-terminal thioester moiety on the other fragment. After rearrangement through an S→N acyl shift, a native peptide bond is formed. The reaction can be performed also in the presence of other unprotected cysteine residues because of a reversible reaction preceding an irreversible step. Using this methodology, a spin-labeled Ras binding domain has been synthesized, showing a stable paramagnetic center detected by EPR (Becker et al., 2005).

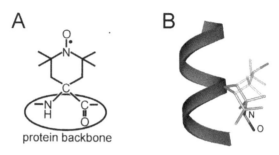

Fig. 2. The TOAC amino acid spin label. (A) chemical structure. (B) three-dimensional structure of the spin label incorporated into an a-helix. The flip of the six-membered ring as the only possible degree of freedom is shown in shaded representation.

2.3 Spin labeling using nonsense suppressor methodology

Spin label amino acids can been introduced into proteins by employing the nonsense suppressor methodology, for example by utilizing the amber suppressor tRNA chemically aminoacylated with the desired spin label amino acid (Cornish et al., 1994). Although this strategy might prove generally applicable in the future using unique transfer RNA(tRNA)/aminoacyl–tRNA-synthetase pairs (Chin et al., 2003), only few laboratories are currently equipped to apply this methodology successfully.

2.4 Spin labeling using nonsense suppressor methodology

As introduced by Kolb, Finn and Sharpless in 2001 (Kolb et al., 2001), the basic concept of "click chemistry" is the highly selective formation of a carbon-heteroatom bond under mild conditions with high yield. Its modular concept renders it a favorable tool for introducing labels into biomolecules. An example is the 1,3-dipolar cycloaddition of organic azides with alkynes in the presence of Cu which has been used to attach fluorescent probes to biomolecules (Deiters and Schultz, 2005). Recently, Tamas and coworkers (Tamas et al., 2009) described the synthesis of nitroxide moieties suitable for click chemistry, thereby opening this approach also for site-directed spin labeling.

3. EPR analysis of spin labeled proteins

In the following, the different experimental techniques of EPR spectroscopy on spin labeled proteins are introduced. Therein, the methods are exemplified with the sensory rhodopsin-transducer complex mediating the photophobic response of the halophilic archaeum

Site-Directed Spin Labeling and Electron Paramagnetic Resonance (EPR) Spectroscopy: A Versatile Tool to Study Protein-Protein Interactions

181

Natronomonas pharaonis. The photophobic response of this organism to green-blue light is mediated by sensory rhodopsin II, *Np*SRII, which is closely related to the light driven proton pump bacteriorhodopsin (for a recent review see (Klare et al., 2007)). *Np*SRII is a seven transmembrane helix (A–G) protein with a retinal chromophore covalently bound via a protonated Schiff base to a conserved lysine residue on helix G. Signal transduction to the intracellular two-component pathway modulating the swimming behavior of the cell takes place via the interaction of *Np*SRII with the tightly bound transducer protein, *Np*HtrII (halobacterial transducer), in a 2:2 complex. A transducer dimer comprising a four-helix transmembrane domain, a linker region consisting of two HAMP domains (Aravind and Ponting, 1999), and a cytoplasmic signaling domain, is flanked by the two SRII receptors.

3.1 Spin label dynamics

The shape of room temperature cw EPR spectra reflects the reorientational motion of the spin label side chain. The influence of spin label dynamics on the spectral shape has been reviewed in detail (Berliner, 1976; Berliner, 1979; Berliner & Reuben, 1989), and the relationship between the dynamics of the spin label side chain and protein structure has been extensively studied for T4 lysozyme (Columbus et al., 2001; Columbus & Hubbell, 2002; Fleissner et al., 2009, 2011; Mchaourab et al., 1996).

In general, the term "mobility" is used to characterize the effects on the EPR spectral features due to the motional rate, amplitude, and anisotropy of the overall reorientational motion of the spin label. Spin labeled sites exposed to the bulk water exhibit weak interaction with the rest of the protein as found for helix surface sites or loop regions and consequently display a high degree of mobility, that is characterized by a small apparent hyperfine splitting and narrow line widths (Fig. 3A & 3B, position 154). If the mobility of the spin label side chain is restricted by interaction with neighboring side chains or backbone atoms of the protein itself or an interaction partner, the line widths and the apparent hyperfine splittings are increased (Fig. 3A & 3B, position 159). Although the relation between the nitroxide dynamics and the EPR spectral line shape is complex, the line width of the centre line, ΔH^0, and the second moment of the spectra, $\langle H^2 \rangle$, have been found to be correlated with the structure of the binding site environment and can therefore be used as mobility parameters (Hubbell et al., 1996; Mchaourab et al., 1996).

The plot of these mobility parameters versus the residue number reveals secondary structure elements through the periodic variation of the mobility as the spin label sequentially samples surface, tertiary, or buried sites. This allows assignment of α-helices, β-strands, or random structures. A more general classification of regions exhibiting buried, surface-exposed, or loop residues can be obtained from the correlation between the inverse of the two mobility parameters, as shown in Figure 3C. Side chains from different topographical regions of a protein can be thereby classified on the basis of the x-ray structures of T4 lysozyme and annexin 12 (Hubbell et al., 1996; Isas et al., 2002; Mchaourab et al., 1996).

For a more quantitative interpretation of the experimental data in terms of dynamic mechanisms and local tertiary interaction, EPR spectra simulations have to be performed. Based on dynamic models developed by Freed and coworkers (Barnes et al., 1999; Borbat et al., 2001; Freed, 1976), excellent agreement of simulations with the corresponding experimental spectra can be obtained. Furthermore, simulations of EPR spectra can be performed on the basis of molecular dynamics (MD) simulations (Beier and Steinhoff, 2006; Budil et al., 2006; Oganesyan, 2007; Sezer et al., 2008; Steinhoff et al., 2000a; Steinhoff and Hubbell, 1996). Thus, a direct link is provided between molecular structure and EPR spectral line shape, thus allowing verification, refinement, or even de-novo prediction (Alexander et al., 2008) of structural models of proteins or protein complexes.

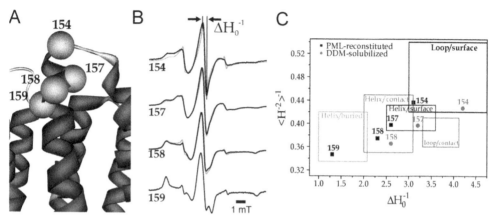

Fig. 3. Mobility analysis of spin labelled proteins. (A) Crystal structure of NpSRII (Luecke et al., 2001). The Cα atoms of spin labeled sites are shown as spheres. (B) X band EPR spectra of spin labeled NpSRII solubilised in detergent (gray) or reconstituted in purple membrane lipids (black). (C) Two-dimensional mobility plot of the inverse of the second moment *vs.* the inverse of the central linewidth (solubilized: gray circles, reconstituted: black squares), determined from the spectra in B. Boxes indicate the topological regions of proteins according to Isas et al. (2002) and Mchaourab et al. (1996).

The motion of a nitroxide spin label side chain is characterized by three correlation times, the rotational correlation time for the entire protein, the effective correlation time due to the rotational isomerization spin label linker, and the effective correlation time for the segmental motion of the protein backbone. These correlation times can significantly differ in the time scales they occur on. Thus, experimental data for all relevant time scales have to be available to set up an appropriate dynamical model. For this case, correlation times from μs (for the overall protein motion) down to ps (for the rotational isomerization) have to be covered by the experiment. EPR spectra at different microwave frequencies are sensitive to motions on different time scales. EPR at lower frequencies is sensitive to slower motions whereas faster motions are completely averaged out. On the other hand, high-frequency EPR can resolve such fast motions, but slower motions are "frozen" at the high-frequency time scale. Consequently, combining experiments at different microwave frequencies (multifrequency EPR) allows separation of various motional modes in a spin labelled protein according to their different time scales. Most of the work so far has been done using spin-labeled T4 lysozyme as a model system (Liang & Freed, 1999; Liang et al., 2004; Zhang et al., 2010).

Proteins and protein complexes are inherently dynamic structures that can exhibit a number of conformational substates often playing a key role for their function (Cooper, 1973,

Site-Directed Spin Labeling and Electron Paramagnetic Resonance (EPR) Spectroscopy: A Versatile Tool to Study
Protein-Protein Interactions

183

Frauenfelder et al., 1988, 1991). A given state of a protein consists of a limited number of such substates with life times in the µs to ms range that can, for example, correspond to "bound" and "unbound" conformations of a protein binding interface. According to the life time of the substates they often can be recognized in room temperature cw spectra of spin labeled proteins if they are characterized by different spin label side chain mobilities due to structural changes in their vicinity. In the past years, Hubbell and co-workers established three experimental techniques to analyze conformational equilibria in proteins and to dissect them from spin label rotameric exchange, namely osmolyte perturbation (Lopez et al., 2009), saturation recovery (Bridges et al., 2010) and high-pressure EPR (McCoy & Hubbell, 2011).

3.2 Spin label solvent accessibilities

Supplementing the motional analysis, the accessibility of the spin label side chain toward paramagnetic probes (exchange reagents), which selectively partition in different environments of the system under investigation, can be used to define the location of spin label with respect to the protein/water/membrane boundaries. The accessibility of the nitroxide spin label side chain is defined by its Heisenberg exchange frequency, W_{ex}, with an exchange reagent diffusing in its environment. Water-soluble metal ion complexes like NiEDDA or chromium oxalate (CrOx) allow to quantify the accessibility from the bulk water phase, whereas molecular oxygen or hydrophobic organic radicals that mainly partition in the hydrophobic part of the lipid bilayer define the accessibility from the lipid phase. The concentration gradients of NiEDDA and molecular oxygen along the membrane normal can be used to characterize the immersion depth of the spin label side chain into the lipid bilayer (Altenbach et al., 1994; Marsh et al., 2006). Two experimental techniques can be used to determine the nitroxide's accessibility toward the paramagnetic probes: Cw power saturation, and saturation recovery.

Most commonly, Heisenberg exchange rates for nitroxide spin label side chains in proteins are measured using cw power saturation. Here, the EPR signal amplitude is monitored as a function of the incident microwave power in the absence and presence of the paramagnetic quencher. From the saturation behaviour of the nitroxide, an accessibility parameter, Π, can be extracted that is proportional to W_{ex} (Altenbach et al., 1989a; Altenbach et al., 2005; Farabakhsh et al., 1992). In Figure 4, the accessibility analysis performed on a 24 amino acid long segment starting at position 78 in the transmembrane region and extending to position 101 in the cytoplasm of the transducer protein NpHtrII in complex NpSRII is shown as an example for this technique (Bordignon et al., 2005). Figure 4A shows the crystal structure of the transmembrane region of the complex (Gordeliy et al., 2002). Power saturation experiments have been performed with air (21% O_2) and 50 mM CrOx, respectively. The Π values calculated from these experiments are shown in panel B versus residue number. The low Π values for both oxygen and CrOx for residues 78 to 86 indicate their location in a densely packed protein-protein interface. The clear periodicity of 3.6 residues (see inset in panel B) reflects the α-helical structure. For positions 87 to 94 a gradual increase in the Π_{CrOx} and Π_{oxygen} values is observed, indicating that this region is protruding away from the protein-protein interface into the cytoplasm. For positions 92 to 101 the Π_{CrOx} and Π_{oxygen} values observed are typical for water exposed residues. Also here a periodical pattern corresponding to an α-helical structure is observed.

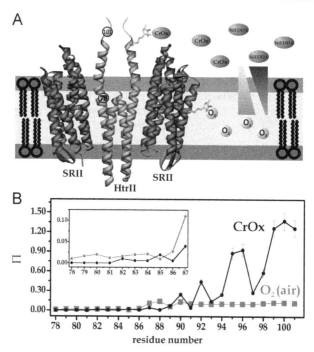

Fig. 4. Spin label accessibilities by power saturation. (A) Structure of the NpSRII/NpHtrII complex in a lipid bilayer (light gray: hydrophobic region, medium gray: headgroup region). The concentration gradients for water-soluble reagents (CrOx and NiEDDA) and lipid-soluble reagents (O_2) are indicated by shaded triangles. The first (78) and last residue (101) of the region investigated are indicated. (B) Accessibility parameters Π_{CrOx} (black circles) and Π_{Oxygen} (gray squares) vs. residue number. Π_{CrOx} values have been obtained with 50 mM CrOx, Π_{Oxygen} values with air (21% O_2). The inset depicts magnified the region from residues 78 to 87 to show the periodicity of 3.6 for Π_{CrOx} and Π_{Oxygen}.

Saturation recovery EPR (SR-EPR) allows measuring the spin–lattice relaxation time T_{1e} which is connected to the Heisenberg exchange frequency W_{ex} by

$$W_{ex} = \Delta \left(\frac{1}{T_{1e}} \right)_R \tag{5}$$

directly. In this type of experiment, saturating microwave pulses are applied to the sample in the absence and in the presence of exchange reagents, and the recovery of the z-magnetization is monitored as a function of time. Analyses of the recovery curves provide T_{1e} and thus the accessibility for the respective exchange reagent (Altenbach et al., 1989a, 1989b; Nielsen et al., 2004). One major advantage of SR-EPR compared to cw power saturation is that in the presence of multiple spin populations (see chapter 2.1) all corresponding T_{1e} values and accessibilities can be determined by SR-EPR. In contrast, cw power saturation can only provide an average accessibility for all components present in the cw EPR spectrum, and moreover, this average value will be biased towards the most mobile

Site-Directed Spin Labeling and Electron Paramagnetic Resonance (EPR) Spectroscopy: A Versatile Tool to Study
Protein-Protein Interactions

185

component as it dominates the amplitude of the resonance lines. Moreover, saturation recovery can be used to distinguish between rotamer exchange for the spin label side chain (~ 0.1-1 µs range) and conformational exchange of the protein, which is at least one order of magnitude slower (Bridges et al., 2010).

3.3 Polarity and proticity of the spin label micro-environment

Polarity and proticity of the spin label microenvironment are reflected in the hyperfine component A_{zz} and the g tensor component g_{xx}. A polar environment shifts A_{zz} to higher values, whereas the tensor component g_{xx}, determined from the B-field of the canonical peak position ($g_{xx} = h\nu / \mu_B B$), is decreased. The A_{zz} component can be obtained from cw X-band EPR spectra of spin labeled proteins in frozen samples. The principal g-tensor components and their variation can be determined with high accuracy using high-field EPR techniques due to the enhanced Zeeman resolution (Steinhoff et al., 2000b). In regular secondary structure elements with anisotropic salvation (e.g., surface exposed α-helices), the water density and hence the tensor component values A_{zz} and g_{xx} are a periodic function of residue number. Therefore, similarly to accessibility measurements with water soluble exchange reagents (see 3.2), these data can be used to obtain structural and topological information and the polarity of the spin label environment can reveal detailed information on the protein fold.

The polarity parameter values for the sequence 88 to 94 in the first HAMP domain of NpHtrII in complex with NpSRII are shown in Figure 5A (Brutlach et al., 2006). It is evident that positions 90 and 93 in NpHtrII are located in a more polar, water accessible environment. The same holds for position 154 on the cytoplasmic surface of NpSRII. In contrast, positions 88, 89, 91, 92 and 94 reside in a more apolar environment that is less accessible to water. Also evident is a periodical pattern, characterizing the α-helical structure. The exceptional apolar character of position 78 indicates that this side chain is deeply buried in a protein-protein or protein-lipid interface. These results are reflected in a structural model, which has been based in

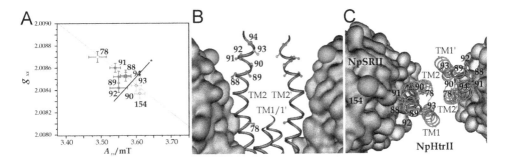

Fig. 5. (A) Plot of g_{xx} versus A_{zz} for positions 88 to 94 in NpHtrII according to Brutlach et al., 2006. The plot also includes values for position 78 on the second transmembrane helix (TM2) and for position 154 in NpSRII. An arbitrary threshold of g_{xx} / A_{zz} indicated by the diagonal line marked with * classifies the sites into more polar (blue) or more apolar sites (red). (B) Side view onto NpSRII (surface representation) and the four-helix bundle of the transducer (ribbon representation) up to position 96 according to Bordignon et al. (2005). (C) Cytoplasmic view of the structural model.

addition on mobility, accessibility and distance data (see 2.4) for this region (Figure 5B and C) (Bordignon et al., 2005). Residues 88, 91 and 92 are located in protein-protein interfaces, and positions 90 and 93 are positioned at the opposite side of the transducer helix. Position 78 is, in line with the exceptional low polarity of its micro-environment, buried in the densely packed four helix bundle of the NpHtrII dimer.

3.4 Inter-spin label distance measurements

If two spin label side chains are introduced into a biomolecule or two singly labeled molecules are in a stable macromolecular complex, the distance between the two labels can be determined through quantification of their spin–spin interaction, thus providing valuable structural information.

Spin–spin interaction is composed of static dipolar interaction, modulation of the dipolar interaction by the residual motion of the spin-label side chains, and exchange interaction. The static dipolar interaction in an unordered immobilized sample leads to considerable broadening of the cw EPR spectrum if the inter spin distance is less than 2 nm (Figure 6A, C). Distances can be quantified by a detailed line shape analysis of EPR spectra of frozen protein samples or proteins in solutions of high viscosity (Altenbach et al., 2001; Rabenstein and Shin, 1995; Steinhoff et al., 1991; Steinhoff et al., 1997). Pulse EPR techniques expand the accessible distance range up to 8 nm (Borbat and Freed, 1999; Pannier et al., 2000; Martin et al., 1998). Two major protocols have been successfully applied, the 4-pulse DEER or 4-pulse PELDOR (Figure 6D) and the Double Quantum Coherence approaches (for a recent review see (Schiemann and Prisner, 2007)).

The combination of cw and pulse EPR techniques, taking into account borderline effects in the region from 1.6-1.9 nm (Banham et al., 2008; Grote et al., 2008), provide means to determine interspin distances in the range from 1-8 nm, thereby covering the most important distance regime necessary for structural investigations on biomacromolecules. Remarkably, the DEER data can provide, besides the distance information, also information about the orientation of the spin label side chains, their conformational flexibility and the spin density distribution, thereby increasing the amount and quality of data for a setup or verification of structural models.

Based on inter spin distance measurements on 26 different pairs of spin labels introduced into the cytoplasmic regions of NpSRII and NpHtrII the arrangement of the transmembrane domains of this complex was modelled (Wegener et al., 2001) (Fig. 7A). Direct comparison of the EPR model with the later determined crystal structure (Gordeliy et al., 2002) (Fig. 7B) shows the consistency of the EPR model with the crystal structure concerning the general topology and the location and relative orientation of the transmembrane helices. Remarkably, also most of the side-chain orientations within the complex coincide quite well in the two models, although for the EPR based model the bacteriorhodopsin structure had to be used as a template for NpSRII, since its structure was not known at that time. In a later study it was shown that the structural properties of the HAMP domain as characterized by mobility, solvent accessibility, and intra-transducer-dimer distance data are in agreement with the NMR model of the HAMP domain from Archaeoglobus fulgidus (Döbber et al., 2008).

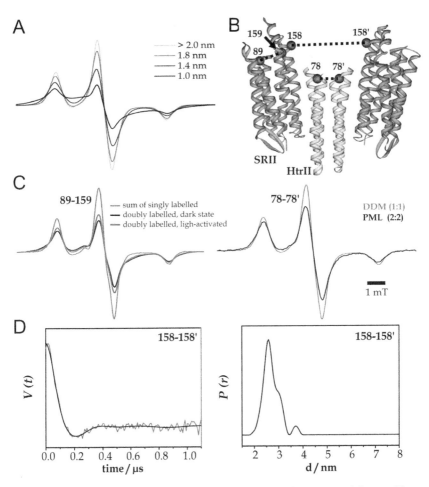

Fig. 6. Interspin distance measurements. (A) Simulated powder spectra (obtained by cw EPR on frozen samples) for different interspin distances. (B) Ribbon representation of the X-ray structure of the 2:2 NpSRII–NpHtrII complex (PDB 1H2S). Positions L89, S158 and L159 in NpSRII, and V78 in NpHtrII are shown as spheres. (C) left: cw EPR spectra (T = 160 K) of the double mutant L89R1/L159R1 in the receptor ground state (black) and in the trapped signaling state (M-state, red) compared to the sum of the spectra of the singly labeled samples (gray) reveals line broadening due to dipolar interaction. Interspin distances of 1.1 (±0.2) nm for the ground state and 1.3 (±0.2) nm for the M-trapped have been determined (Bordignon et al., 2007); right: NpHtrII–V78R1 solubilized in DDM (gray) or reconstituted in PML (black) in the absence of NpSRII. The interspin distance obtained in the reconstituted sample is 1.3 (±0.2) nm (Klare et al., 2006). (D) Left: background corrected DEER time domain for NpSRII-S158R1. The distance distribution shown in the right panel shows a mean distance of 2.6 nm between the two spin labels bound to positions 158 in the 2:2 complex that is in good agreement with the distance of 3 nm between the oxygen atoms of the respective serine residues as calculated from the crystal structure.

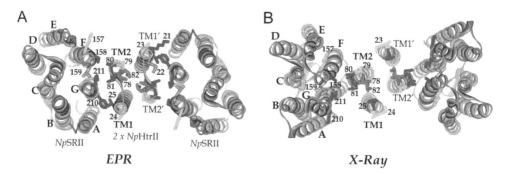

Fig. 7. (A) EPR based model of the transmembrane region of the NpSRII/NpHtrII complex (viewed from the cytoplasmic side). Side chains where spin labels have been attached are shown in stick representation. The color code represents the strength of the observed dipolar interaction (blue and red: strong; cyan and orange: weak). (B) Crystal structure of NpSRII/NpHtrII (PDB: 1H2S) (Klare et al., 2004). The color code for the side chains is the same as in (A).

4. SDSL EPR in protein interaction studies

4.1 Structure of the Na⁺/H⁺ antiporter dimer

An excellent example for the application of DEER spectroscopy to investigate interactions between proteins is the *E. coli* Na^+/H^+ Antiporter NhaA (Hilger et al., 2005, 2007). The protein is responsible for the specific exchange of Na^+ for H^+, and is known to be regulated by pH. From studies on two-dimensional crystals diffracting at 4 Å it had been revealed that NhaA forms a dimer in crystals, and *in vivo* studies as well as cross-linking data suggested that it also works as a dimer *in vivo*. Applying DEER on a NhaA variant labeled at position H225, it could be shown (Hilger et al., 2005) that a pH independent distance of 4.4 nm between spin labeled sites in neighboring molecules can be resolved, and that the degree of dimerization, as judged from the modulation depth of the DEER dipolar evolution data, strongly depends on pH. Thereby, experiments utilizing a singly spin labelled position yielded data strongly supporting the stoichiometry of the functional unit and providing evidences for a mechanistic picture of pH regulation, i.e. that the affinity between the protomers is modulated.

In a later study (Hilger et al., 2007), an extensive distance mapping of the NhaA dimer with nine different spin labeled amino acid positions was carried out. Based on these distance distribution data, explicit modeling of the spin label side chain conformations with a rotamer library approach, and combination with the available X-ray structures, a high-resolution structure of the presumably physiological dimer was determined.

4.2 Structure and function of the tRNA modifying MnmE/GidA complex

The GTP hydrolyzing protein MnmE is, together with the protein GidA, involved in the modification of the wobble position of certain tRNAs (Meyer et al, 2009). It belongs to the expanding class of G proteins activated by nucleotide-dependent dimerization (GADs). The crystal structure shows that MnmE is a multidomain protein with a central helical part in which a Ras-like domain is inserted, and an N-terminal tetrahydrofolate-binding unit. MnmE was predicted to form a dimer in solution, in which the two G domains are separated with a distance of about 50Å between the two P-loops (Fig. 10A). A G domain dimerization had been proposed based on biochemical data and on the crystal structure of the isolated G domains in complex with the GTP hydrolysis transition state mimic GDP·AlF$_x$.

In a DEER study, distance measurements between spin labels positioned in the MnmE G domain and in the dimerization domain have been carried out (Meyer et al., 2009).The distance distributions for position E287 in the G domains (Fig. 8A) are shown in Figure 8B. In the apo state without any nucleotide bound, the two spin labels exhibit a distances of 55Å for E287R1 and a broad distribution of distances from 25Å to 50Å. GTP (GppNHp) binding induces the formation of additional distances at 27Å (S278R1) / 37Å (E287R1), contributing about 30% to the distance distributions. Upon GTP hydrolysis (GDP · AlF$_x$ state), the distance distribution shows a single population maximum at 28Å (S278R1) / 36Å (E287R1). Thus, an equilibrium between an open conformation with distant G domains and a closed conformation with the G domains being in close proximity, exists when GTP is bound, and is shifted completely towards the closed state upon hydrolysis. A schematic representation of the proposed conformational changes of the MnmE G domains based on the EPR data is depicted in Figure 8C.

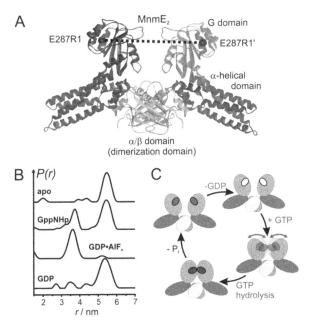

Fig. 8. G domain dimerization of MnmE monitored by DEER. (A) Structural model of the MnmE-dimer. Position 287 in the G domains, which was spin labeled is indicated by black spheres. (B) DEER distance distributions obtained by Tikhonov regularization using the program DeerAnalysis 2008. (C) Schematic representation of the G domain conformational states during the GTPase cycle.

In addition, spin labels attached at position 105 in the dimerization domain showed no significant distance changes during the GTPase cycle, indicating that the initial dimerization interface is largely preserved despite the large G domain movements. In addition, a dependency of the GTPase activity and consequently of the G domain motions on the presence of specific cations could be fully corroborated by the DEER analysis performed in this study. In a subsequent study, the influence of binding of GidA to MnmE in a 2:2 complex was investigated, showing that the interaction of GidA with MnmE partly abolishes the previously observed cation dependency.

4.3 Subunit binding in the chaperone/usher pathway of pilus biogenesis

Type 1 pili are adhesive multisubunit fibres in Gram-negative bacteria. During pilus assembly, subunits dock as chaperone-bound complexes to an usher, which catalyses their polymerization and pilus translocation across the membrane. In the background of the recently determined crystal structure of the full-length FimD usher bound to the FimC–FimH chaperone–adhesin complex, SDSL EPR was used to show that subsequent subunits bind to the usher c-terminal domains after undergoing so-called donor-strand exchange (Phan et al., 2011).

DEER spectroscopy was carried out on spin label pairs introduced into Fim proteins and the results ware compared to calculated distance distributions obtained from alternative models of the complex. The pair residue 74 of FimC and residue 756 of FimD in the complex revealed that the distance distribution obtained experimentally overlaps with that predicted when FimC–FimG (FimG is the adjacent subunit within the pilus) locates at the c-terminal domains. For the FimC-Q74R1/FimD-S774R1 the experimental distance distributions compared with those calculated for the crystal structure of FimD-C-H, assuming that the position of FimC-G, is similar to the previously bound chaperone-subunit complex FimC-H, are in good agreement, further supporting the assumption that FimC-G in the FimD-C-G-H complex locates at the c-terminal domains and is bound to c-terminal domain 2 (Phan et al., 2011).

4.4 Conformation of peptides bound to TAP

The ATP-binding cassette transporter associated with antigen processing (TAP) is involved in the adaptive immune defense against infected or malignantly transformed cells. It translocates proteasomal degradation products, i.e. peptides of 8 to 40 residues, into the lumen of the endoplasmic reticulum for loading onto MHC class I molecules. EPR spectroscopy and simulations based on a rotamer library were used to reveal conformational details about the bound peptides (Herget et al., 2011).

The authors used two different spin label side chains, namely the PROXYL spin label to monitor side-chain dynamics and environmental, and TOAC-labeled peptides (see chapter 2.2) to detect backbone properties. For different locations of the spin label on the peptide, striking differences in affinity, dynamics, and polarity were found. The mobility of the spin labels was found to be strongly restricted at the ends of the peptide. In contrast, the central region was flexible, suggesting a central peptide bulge. Furthermore, DEER spectroscopy was used for the determination of intrapeptide distances in doubly labeled peptides bound to TAP. Comparison with calculated distance distributions based on a rotamer library led

Site-Directed Spin Labeling and Electron Paramagnetic Resonance (EPR) Spectroscopy: A Versatile Tool to Study
Protein-Protein Interactions

191

the authors to the conclusion that peptides bind to TAP in an extended kinked structure, analogous to those bound to MHC class I proteins (Herget et al., 2011).

5. Acknowledgement

Part of this work was supported by the Deutsche Forschungsgemeinschaft.

6. References

Alexander, N., Al-Mestarihi, A., Bortolus, M., Mchaourab, H.S. & Meiler, J. (2008) De Novo High-Resolution Protein Structure Determination from Sparse Spin-Labeling EPR Data, *Structure*, Vol. 16, No. 2, (February 2008), pp. 181-195, ISSN 0969-2126

Altenbach, C., Flitsch, S.L., Khorana, H.G. & Hubbell, W.L. (1989a). Structural studies on transmembrane proteins, 2: spin labeling of bacteriorhodopsin mutants at unique cysteines. *Biochemistry*, Vol. 28, No. 19, (September 1989), pp. 7806-7812, ISSN 0006-2960

Altenbach, C., Froncisz, W., Hyde, J.S. & Hubbell, W.L. (1989b). Conformation of spin-labeled melittin at membrane surfaces investigated by pulse saturation recovery and continuous wave power saturation electron-paramagnetic resonance. *Biophysical Journal*, Vol. 56, No. 6, (December 1989), pp. 1183-1191, ISSN 1542-0086

Altenbach, C., Marti, T., Khorana, H.G. & Hubbell, W.L. (1990). Transmembrane protein structure: spin labeling of bacteriorhodopsin mutants. *Science*, Vol. 248, No. 4959, (June 1990), pp. 1088-1092, ISSN 1095-9203

Altenbach, C., Greenhalgh, D.A., Khorana, H.G. & Hubbell, W.L. (1994). A collision gradient method to determine the immersion depth of nitroxides in lipid bilayers: application to spin-labeled mutants of bacteriorhodopsin. *Proceedings of the National Academy of Sciences of the USA*, Vol. 91, No. 5, (March 1994), pp. 1667-1671, ISSN 1091-6490

Altenbach, C., Oh, K.J., Trabanino, R.J., Hideg, K. & Hubbell,W.L. (2001). Estimation of Inter-Residue Distances in Spin Labeled Proteins at Physiological Temperatures: Experimental Strategies and Practical Limitations. *Biochemistry*, Vol. 40, No. 51, (December 2011), pp. 15471-15482, ISSN 0006-2960

Altenbach, C., Froncisz, W., Hemker, R., Mchaourab, H.S. & Hubbell, W.L. (2005). Accessibility of nitroxide side chains: absolute Heisenberg exchange rates from power saturation EPR. *Biophysical Journal*, Vol. 89, No. 3, (September 2005), pp. 2103-2112, ISSN 1542-0086

Anderson,D.J., Hanson,P., McNulty,J., Millhauser,G.L., Monaco,V., Formaggio,F. et al. (1999). Solution Structures of TOAC-Labeled Trichogin GA IV Peptides from Allowed (g = 2) and Half-Field Electron Spin Resonance. *Journal of the American Chemical Society*, Vol. 121, No. 29, (July 1999), pp. 6919-6927, ISSN 1520-5126

Aravind, L. & Ponting, C.P. (1999). The cytoplasmic helical linker domain of receptor histidine kinase and methyl-accepting proteins is common to many prokaryotic signalling proteins. *FEMS Microbiology Letters*, Vol. 176, No. 1, (July 1999), pp. 111-116, ISSN 0378-1097

Banham, J.E., Baker, C.M., Ceola, S., Day, I.J., Grant, G.H., Groenen, E.J.J., Rodgers, C.T., Jeschke, G. & Timmel, C.R. (2008). Distance measurements in the borderline region of applicability of CW EPR and DEER: A model study on a homologous series of spin-labelled peptides. *Journal of Magnetic Resonance*, Vol. 191, No. 2, (April 2008), pp. 202-218, ISSN 1090-7807

Barbosa,S.R., Cilli,E.M., Lamy-Freund,M.T., Castrucci,A.M., and Nakaie,C.R. (1999). First synthesis of a fully active spin-labeled peptide hormone. *FEBS Letters*, Vol. 446, No. 1, (March 1999), pp. 45-48, ISSN 0014-5793

Becker, C.F.W., Hunter, C.L., Seidel, R., Kent, .B.H., Goody, R.S. & Engelhard,M. (2003). Total chemical synthesis of a functional interacting protein pair: The protooncogene H-Ras and the Ras-binding domain of its effector c-Raf1. *Proceedings of the National Academy of Sciences of the USA*, Vol. 100, No. 9, (April 2003), pp. 5075-5080, ISSN 1091-6490

Becker, C.F.W., Lausecker, K., Balog, M., Kalai, T., Hideg, K., Steinhoff, H.-J. & Engelhard,M. (2005). Incorporation of spin-labelled amino acids into proteins. Magnetic Resonance in Chemistry, Vol. 43, No. S1, (December 2005), pp. S34-S39, ISSN 1097-458X

Beier, C. & Steinhoff, H.-J. (2006). A structure-based simulation approach for electron paramagnetic resonance spectra using molecular and stochastic dynamics simulations. *Biophysical Journal*, Vol. 91, No. 7, (October 2006), pp. 2647-2664, ISSN 1542-0086

Berliner, L.J., (Ed.). (June 1976). *Spin labeling: theory and applications*. Academic Press, ISBN 978-0120923502, New York

Berliner, L.J., (Ed.). (September 1979). *Spin labeling II: theory and applications*. Academic Press, ISBN 978-0120923526, New York

Berliner,L.J., Grunwald,J., Hankovszky,H.O., and Hideg,K. (1982). A novel reversible thiol-specific spin label: Papain active site labeling and inhibition. *Analytical Biochemistry*, Vol. 119, No. 2, (January 1982), pp. 450-455, ISSN 0003-2697

Berliner, L.J. & Reuben, J., (Eds.). (June 1989). *Spin labeling*, Vol. 8: *Biological magnetic resonance*. Plenum Press, ISBN 978-0306430725, New York

Borbat, P.P. & Freed, J.H. (1999). Multiple-quantum ESR and distance measurements. *Chemical Physics Letters*, Vol. 313, No. 1-2, (November 1999), pp. 145-154, ISSN 0009-2614

Bordignon, E., Klare, J.P., Döbber, M.A., Wegener, A.A., Martell, S., Engelhard, M. & Steinhoff, H.-J. (2005). Structural Analysis of a HAMP domain: The Linker Region of the Phototransducer in Complex with Sensory Rhodopsin II. *Journal of Biological Chemistry*, Vol. 280, No. 46, (November 2005), pp. 38767-38775, ISSN 1083-351X

Bordignon, E. & Steinhoff, H.-J. (February 2007). Membrane protein structure and dynamics studied by site-directed spin labeling ESR., In: *ESR spectroscopy in membrane biophysics*, Hemminga, M.A. & Berliner, L.J. (Eds.), pp. 129-164, Springer Science and Business Media, ISBN 978-0387250663, New York

Bridges, M.D., Hideg, K. & Hubbell, W.L. (2010). Resolving Conformational and Rotameric Exchange in Spin-Labeled Proteins Using Saturation Recovery EPR. *Applied Magnetic Resonance*, Vol. 37, No. 1, (January 2010), pp. 363-390, ISSN 0937-9347

Brutlach, H., Bordignon, E., Urban, L., Klare, J.P., Reyher, H.-J., Engelhard, M. & Steinhoff, H.-J. (2006). High-Field EPR and Site-Directed Spin Labeling Reveal a Periodical Polarity Profile: The Sequence 88 to 94 of the Phototransducer, NpHtrII, in Complex with Sensory Rhodopsin, NpSRII. *Applied Magnetic Resonance*, Vol. 30, No. 3-4, (June 2006), pp. 359-372, ISSN 0937-9347

Site-Directed Spin Labeling and Electron Paramagnetic Resonance (EPR) Spectroscopy: A Versatile Tool to Study
Protein-Protein Interactions

193

Budil, D.E., Sale, K.L., Khairy, K. & Fajer, P.G. (2006). Calculating slow-motional electron paramagnetic resonance spectra from molecular dynamics using a diffusion operator approach. *Journal of Physical Chemistry A*, Vol. 110, No. 10, (March 2006), pp. 3703-3713, ISSN 1520-5215

Chin, J.W., Cropp, T.A., Anderson, J.C., Mukherji, M., Zhang, Z. & Schultz,P.G. (2003). An Expanded Eukaryotic Genetic Code. *Science*, Vol. 301, No. 5635, (August 2003), pp. 964-967, ISSN 1095-9203

Columbus, L., Kalai, T., Jekö, J., Hideg, K. & Hubbell, W.L. (2001). Molecular motion of spin labelled side chains in α-helices: analysis by variation of side chain structure. *Biochemistry*, Vol. 40, No. 13, (April 2001), pp. 3828-3846, ISSN 0006-2960

Columbus, L. & Hubbell, W.L. (2002). A new spin on protein dynamics. *Trends in Biochemical Sciences*, Vol. 27, No. 6, (June 2002), pp. 288-295, ISSN 0968-0004

Cooper, A. (1976). Thermodynamic fluctuations in protein molecules. *Proceedings of the National Academy of Sciences of the USA*, Vol. 73, No. 8, (August 1976), pp. 2740-2741, ISSN 1091-6490

Cornish, V.W., Benson, D.R., Altenbach, C., Hideg, K., Hubbell, W.L. & Schultz, P.G. (1994). Site-Specific incorporation of biophysical probes into proteins. *Proceedings of the National Academy of Sciences of the USA*, Vol. 91, No. 8, (April 1994), pp. 2910-2914, ISSN 1091-6490

Deiters, A. & Schultz, P. G. (2005). In vivo incorporation of an alkyne into proteins in Escherichia coli. *Bioorganic & Medicinal Chemistry Letters*, Vol. 15, No. 5, (March 2005), pp. 1521-1524, ISSN 0960-894X

Elsässer,C., Monien,B., Haehnel,W. & Bittl,R. (2005). Orientation of spin labels in de novo peptides. Magnetic Resonance in Chemistry, Vol. 43, No. S1, (December 2005), pp. S26-S33, ISSN 1097-458X

Farahbakhsh, Z.Z., Altenbach, C. & Hubbell, W.L. (1992). Spin labeled cysteines as sensors for protein lipid interaction and conformation in rhodopsin. *Photochemistry and Photobiology*, Vol. 56, No. 6, (December 1992), pp. 1019-1033, ISSN 1751-1097

Fleissner, M.R., Cascio, D. & Hubbell, W.L. (2009). Structural origin of weakly ordered nitroxide motion in spin-labeled proteins. *Protein Science*, Vol. 18, No. 5, (May 2009), pp. 893-908, ISSN 1469-896X

Fleissner, M.R., Bridges, M.D., Brooks, E.K., Cascio, D., Kalai, T., Hideg, K. & Hubbell, W.L. (2011). Structure and dynamics of a conformationally constrained nitroxide side chain and applications in EPR spectroscopy. *Proceedings of the National Academy of Sciences of the USA*, Vol. 108, No. 39, (September 2011), pp. 16241-16246, ISSN 1091-6490

Frauenfelder, H., Parak, F.G. & Young, R.D. (1988). Conformational Substates in Proteins. *Annual Reviews of Biophysics and Biophysical Chemistry*, Vol. 17, No. 1, (June 1988), pp. 451-479, ISSN 0084-6589

Frauenfelder, H., Sligar, S.G. & Wolynes, P.G. (1991). The energy landscapes and motions of proteins. *Science*, Vol. 254, No. 5038, (December 1991), pp. 1598-1603, ISSN 1095-9203

Griffith,O.H., and McConnell,H.M. (1966). A Nitroxide-Maleimide Spin Label. *Proceedings of the National Academy of Sciences of the USA*, Vol. 55, No. 1, (January 1966), pp. 8-11, ISSN 1091-6490

Grote, M., Bordignon, E., Polyhach, Y., Jeschke, G., Steinhoff, H.-J. & Schneider, E. (2008). A Comparative EPR Study of the Nucleotide-binding Domains' Catalytic Cycle in the

Assembled Maltose ABC-Importer. *Biophysical Journal*, Vol. 95, No. 6, (June 2008), pp. 2924-2938, ISSN 1542-0086

Hanson,P., Millhauser,G., Formaggio,F., Crisma,M. & Toniolo, C. (1996) ESR Characterization of Hexameric, Helical Peptides Using Double TOAC Spin Labeling. *Journal of the American Chemical Society*, Vol. 118, No. 32, (August 1996), pp. 7618-7625, ISSN 1520-5126

Herget, M., Baldauf, C., Schoelz, C., Parcej, D., Wiesmueller, K.-H., Tampe, R., Abele, R. & Bordignon, E. (2011). Conformation of peptides bound to the transporter associated with antigen processing (TAP). *Proceedings of the National Academy of Sciences of the USA* , Vol. 108, No. 4, (January 2011), pp. 1349-1354, ISSN 1091-6490

Hilger, D., Jung, H., Padan, E., Wegener, C., Vogel, K.P., Steinhoff, H.-J. & Jeschke, G. (2005). Assessing Oligomerization of Membrane Proteins by Four-Pulse DEER: pH-Dependent Dimerization of NhaA Na+/H+ Antiporter of E. coli, *Biophysical Journal*, Vol. 89, No. 2, (August 2005), pp. 1328-1338, ISSN 1542-0086

Hilger, D., Polyhach, Y., Padan, E., Jung, H. & Jeschke, G. (2007). High-Resolution Structure of a Na+/H+ Antiporter Dimer Obtained by Pulsed Electron Paramagnetic Resonance Distance Measurements, *Biophysical Journal*, Vol. 93, No. 10, (November 2007), pp. 3675-3683, ISSN 1542-0086

Hubbell, W.L., Mchaourab, H.S., Altenbach, C., & Lietzow, M.A. (1996). Watching proteins move using site-directed spin labeling. *Structure*, Vol. 4, No. 7, (July 1996), pp. 779-783, ISSN 0969-2126

Hubbell, W.L., Gross, A., Langen, R. & Lietzow, M.A. (1998). Recent advances in site-directed spin labeling of proteins. *Current Opinion in Structural Biology*, Vol. 8, No. 5, (October 1998), pp. 649-656, ISSN 0959-440X

Isas, J.M., Langen, R., Haigler, H.T. & Hubbell, W.L. (2002). Structure and dynamics of a helical hairpin and loop region in annexin 12: a site-directed spin labeling study. *Biochemistry*, Vol. 41, No. 5, (February 2002), pp. 1464-1473, ISSN 0006-2960

Kalai, T., Hubbell, W.L. & Hideg, K. (2009). Click Reactions with Nitroxides, *Synthesis*, Vol. 2009, No. 8, (April 2009), pp. 1336-1340, ISSN 1437-210X

Klare, J.P., Bordignon, E., Engelhard, M. & Steinhoff, H.-J. (2004). Sensory rhodopsin II and bacteriorhodopsin, light activated helix F movement. *Photochemical and Photobiolological Sciences*, Vol. 3, No. 6, (June 2004), pp. 543–547, ISSN 1474-9092

Klare, J.P., Bordignon, E., Döbber, M.A., Fitter, J., Kriegsmann, J., Chizhov, I. et al. (2006). Effects of solubilization on the structure and function of the sensory rhodopsin II/transducer complex. *Journal of Molecular Biology*, Vol. 356, No. 5, (March 2006), pp. 1207–1221, ISSN 0022-2836

Klare, J.P., Chizhov, I. & Engelhard,M. (2007). Microbial Rhodopsins: Scaffolds for Ion Pumps, Channels, and Sensors. Results and Problems in Cell Differentiation, Vol. 45, pp. 73-122, ISSN 1861-0412

Klare, J.P. & Steinhoff, H.-J. (2009). Spin labeling EPR. *Photosynthesis Research*, Vol. 102, No. 2-3, (December 2009), pp. 377-390, ISSN 0166-8595

Klug, C.S. & Feix, J.B. (November 2007). Methods and applications of site-directed spin labeling EPR spectroscopy., In: *Methods in cell biology. Biophysical tools for biologists, volume one: in vitro techniques*, Correia, J.J. & Detrich, H.W. (Eds.), pp. 617-658, Academic Press, ISBN 978-0123725202, New York

Kolb, H., Finn, M. G. & Sharpless, K. B. (2001). Click Chemistry: Diverse Chemical Function from a Few Good Reactions, *Angewandte Chemie International Edition*, Vol. 40, No. 11, (June 2001), pp. 2004-2021, ISSN 1521-3773

Liang, Z. & Freed, J.H. (1999). An assessment of the Applicability of Multifrequency ESR to Study the Complex Dynamics of Biomolecules. *Journal of Physical Chemistry B*, Vol. 103, No. 30, (July 1999), pp. 6384-6396, ISSN 1520-6106

Liang, Z., Lou, Y., Freed, J.H., Columbus, L. & Hubbell, W.L. (2004). A Multifrequency Electron Spin Resonance Study of T4 Lysozyme Dynamics Using the Slowly Relaxing Local Structure Model. *Journal of Physical Chemistry B*, Vol. 108, No. 45, (November 2004), pp. 17649-17659, ISSN 1520-6106

Lopez, C.J., Fleissner, M.R., Guo, Z., Kusnetzow, A.N. & Hubbell, W.L. (2009). Osmolyte perturbation reveals conformational equilibria in spin-labeled proteins. *Protein Science*, Vol. 18, No. 8, (June 2009), pp. 1637-1652, ISSN 1469-896X

Luecke, H., Schobert, B., Lanyi, J.K., Spudich, E.N. & Spudich, J.L. (2001). Crystal structure of sensory rhodopsin II at 2.4 Å: insights into color tuning and transducer interaction. *Science*, Vol. 293, No. 5534, (August 2001), pp. 1499-1503, ISSN 1095-9203

Marsh, D., Dzikovski, B.G. & Livshits, V.A. (2006). Oxygen profiles in membranes. *Biophysical Journal*, Vol. 90, No. 7, (April 2006), pp. L49-L51, ISSN 1542-0086

Marsh, D., Jost, M., Peggion, & Toniolo, C. (2007). TOAC Spin Labels in the Backbone of Alamethicin: EPR Studies in Lipid Membranes. Biophysical Journal, Vol. 92, No. 2, (January 2007), pp. 473-481, ISSN 1542-0086

Martin, R. E., Pannier, M., Diederich, F., Gramlich, V., Hubrich, M. & Spiess, H. W. (1998). Determination of End-to-End Distances in a Series of TEMPO Diradicals of up to 2.8 nm Length with a New Four-Pulse Double Electron Electron Resonance Experiment, *Angewandte Chemie International Edition*, Vol. 37, No. 20, (December 1998), pp. 2833-2837, ISSN 1521-3773

Mchaourab, H.S., Lietzow, M.A., Hideg, K. & Hubbell, W.L. (1996). Motion of spin-labeled side chains in T4 lysozyme. Correlation with protein structure and dynamics. *Biochemistry*, Vol. 35, No. 24, (June 1996), pp. 7692-7704, ISSN 0006-2960

McCoy, J. & Hubbell, W.L. (2011). High-pressure EPR reveals conformational equilibria and volumetric properties of spin-labeled proteins. *Proceedings of the National Academy of Sciences of the USA*, Vol. 108, No. 4, (January 2011), pp. 1331-1336, ISSN 1091-6490

Merrifield, B. (1963). Solid Phase Peptide Synthesis. I. The Synthesis of a Tetrapeptide. *Journal of the American Chemical Society*, Vol. 85, No. 14, (July 1963), pp. 2149-2154, ISSN 1520-5126

Nielsen, R.D., Canaan, S., Gladden, J.A., Gelb, M.H., Mailer, C. & Robinson, B.H. (2004). Comparing continuous wave progressive power saturation EPR and time domain saturation recovery EPR over the entire motional range of nitroxides psin labels. *Journal of Magnetic Resonance*, Vol. 169, No. 1, (July 2004), pp. 129-163, ISSN 1090-7807

Oganesyan, V.S. (2007). A novel approach to the simulation of nitroxide spin label EPR sdpectra from a single truncated dynamical trajectory. *Journal of Magnetic Resonance*, Vol. 188, No. 2, (October 2007), pp. 196-205, ISSN 1090-7807

Pannier, M., Veit, S., Godt, A., Jeschke, G. & Spiess, H.W. (2000). Dead-Time Free Measurement of Dipole-Dipole Interactions between Electron Spins. *Journal of Magnetic Resonance*, Vol. 142, No. 2, (February 2000), pp. 331-340, ISSN 1090-7807

Phan, G., Remaut, H., Wang, T., Allen, W.J., Pirker, K.F. et al. (2011). Crystal structure of the FiumD usher bound to its cognate FimC-FimH substrate. *Nature*, Vol. 474, No. 7349, (June 2011), pp. 49-53, ISSN 0028-0836

Rabenstein, M.D. & Shin, Y.K. (1995). Determination of the distance between 2 spin labels attached to a macromolecule. *Proceedings of the National Academy of Sciences of the USA*, Vol. 92, No. 18, (August 1995), pp. 8239–8243, ISSN 1091-6490

Rassat,A. & Rey, P. (1967). Nitroxides, 23: preparation of amino-acid free radicals and their complex salts. *Bulletin De La Societe Chimique De France*, Vol. 3, (March 1967), pp. 815-818, ISSN 0037-8968

Schiemann, O. & Prisner, T.F. (2007). Long-range distance determinations in biomacromolecules by EPR spectroscopy. *Quarterly Reviews of Biophysics*, Vol. 40, No. 1, (June 2007), pp. 1-53, ISSN 1469-8994

Sezer, D., Freed, J.H. & Roux, B. (2008). Simulating electron spin resonance spectra of nitroxide spin labels from molecular dynamics and stochastic trajectories. *Journal of Chemical Physics*, Vol. 128, No. 16, (April 2008), pp. 165106-165116, ISSN 1089-7690

Smirnov,A.F., Ruuge,A., Reznikov,V.A., Voinov,M.A., and Grigor'ev,I.A. (2004). Site-Directed Electrostatic Measurements with a Thiol-Specific pH-Sensitive Nitroxide: Differentiating Local pK and Polarity Effects by High-Field EPR. *Journal of the American Chemical Society*, Vol. 126, No. 29, (July 2004), pp. 8872-8873, ISSN 1520-5126

Steinhoff, H.-J., Dombrowsky, O., Karim, C., & Schneiderhahn, C. (1991). Two dimensional diffusion of small molecules on protein surfaces: an EPR study of the restricted translational diffusion of protein-bound spin labels. *European Biophysics Journal*, Vol. 20, No. 5, (December 1991), pp. 293-303, ISSN 0175-7571

Steinhoff, H.-J. & Hubbell, W.L. (1996). Calculation of electron paramagnetic resonance spectra from Brownian dynamics trajectories: application to nitroxide side chains in proteins. *Biophysical Journal*, Vol. 71, No. 4, (October 1996), pp. 2201-2212, ISSN 1542-0086

Steinhoff, H.-J., Radzwill, N., Thevis, W., Lenz, V., Brandenburg, D., Antson, A., Dodson, G.G. & Wollmer, A. (1997). Determination of interspin distances between spin labels attached to insulin: comparison of electron paramagnetic resonance data with the X- ray structure. *Biophysical Journal*, Vol. 73, No. 6, (December 1997), pp. 3287-3298, ISSN 1542-0086

Steinhoff, H.-J., Müller, M., Beier, C. & Pfeiffer, M. (2000a). Molecular dynamics simulation and EPR spectroscopy of nitroxide side chains in bacteriorhodopsin. *Journal of Molecular Liquids*, Vol. 84, No. 1, (January 2000), pp. 17-27, ISSN 0167-7322

Steinhoff, H.J., Savitsky, A., Wegener, C., Pfeiffer, M., Plato, M. & Möbius, K. (2000b). High-field EPR studies of the structure and conformational changes of site-directed spin labeled bacteriorhodopsin. *Biochimica et Biophysica Acta*, Vol. 1457, No. 3, (April 2000), pp.253–262, ISSN 0005-2728

Wegener, A.A., Klare, J.P., Engelhard, M. & Steinhoff, H.-J. (2001). Structural insights into the early steps of receptor-transducer signal transfer in archaeal phototaxis. *EMBO Journal*, Vol. 20, No. 19, (October 2001), pp. 5312-5319, ISSN 1460-2075

Zhang, Z., Fleissner, M.R., Tipikin, D.S., Liang, Z., Moscicki, J.K., Earle, K.A., Hubbell, W.L. & Freed, J.H. (2010). Multifrequency Electron Spin Resonance Study of the Dynamics of Spin Labeled T4 Lysozyme. *Journal of Physical Chemistry B*, Vol. 114, No. 16, (April 2010), pp. 5503-5521, ISSN 1520-6106

10

Modification, Development, Application and Prospects of Tandem Affinity Purification Method

Xiaoli Xu[1,2,*], Xueyong Li[1,*],
Hua Zhang[2] and Lizhe An[2]
[1]Department of Burn and Plastic Surgery, Tangdu Hospital,
Fourth Military Medical University, Xi'an,
[2]Key Laboratory of Arid and Grassland Agroecology of Ministry of Education,
School of Life Sciences, Lanzhou University, Lanzhou,
[1,2]China

1. Introduction

Since the completion of genome sequences of several organisms, attention has been focused on the analysis of the function and functional network of proteins. Most cell type-specific functions and phenotypes are mediated and regulated by the activities of multiprotein complexes as well as other types of protein–protein interactions and posttranslational modifications. Accordingly, the formation and function of macromolecular protein complexes support the whole of cell processes. Consequently, analysis of the variations of protein complex composition in different cell and tissue types is essential to understand the relationship between gene products and cellular functions in diverse physiological contexts (Alberts, 1998; Cusick et al., 2005).

With the development of research strategies, many large-scaleprotein-protein interaction studies have been performed in model organisms, especially the budding yeast Saccharomyces cervisiae. Genome-wide yeast two-hybrid screens (Fromont-Racine et al., 1997; Ito et al., 2001; Uetz et al., 2000) and protein chip-based methods (Zhu et al., 2001) allow broader insight into the interaction networks and afford the possibility of high-throughput analysis of function and functional network of proteins. While the former approach provides information relating to interactions between two proteins, typically of binary nature, and has the potential for false-positive and false-negative results, the latter approach is time consuming and labor intensive. These defects may limit their application in large scale protein complex purification.

A novel protein complex purification strategy, named tandem affinity purification (TAP) (Puig et al., 2001; Rigaut et al., 1999), in cooperation with mass spectrometry allows identification of interaction partners and purification of protein complexes. This strategy was originally developed in yeast and has been tested in many cells and organisms.

---

* These authors contributed equally to this work

2. TAP method: A brief overview

The basic principle of TAP is similar to the epitope tagging strategy but different on the utilization of two sequential tags instead of one. Rigaut et al. (Rigaut et al., 1999) compared several tag combinations aiming at high recovery rates without hampering protein functions and developed the standard TAP tag. The TAP method requires fusing a TAP tag to the target protein. The TAP tag consist of two IgG-binding domains of protein A of *Staphylococcus aureus* (ProtA) and a calmodulin-binding domain (CBP), separated by a cleavage site for the tobacco etch virus (TEV) protease (Rigaut et al., 1999). In addition to the C-terminal TAP tag, an N- terminal TAP tag (Puig et al., 2001), which is a reverse orientation of the C-terminal TAP tag, was also generated (Fig. 1A).

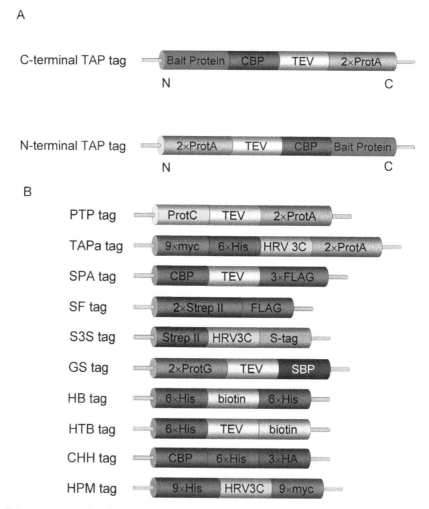

Fig. 1. Diagrammatic sketch of the TAP tag. (A) The original C- and N-terminal TAP tag. (B) Variation of TAP tags developed over the past few years.

The TAP method requires the fusion of the TAP tag to proteins of interest, either at the C- or N-terminus, and the transformation of the construct into appropriate host organisms. The TAP-tagged protein is expressed in host cells at close to physiological concentrations to form a complex with endogenous components. Extracts prepared from cells expressing TAP-tagged proteins are subjected to two sequential purification steps (Fig. 2).

It is well known that the TAP system is very useful for the identification of relatively stable protein complexes, and have helped in the discovery of novel interactions. The TAP method has many advantages: first, the TAP system allows rapid purification of protein complexes without the knowledge of their function or structure. Second, the TAP method enables protein complex purification under native conditions. Third, the tandem purification steps provide highly specific and reduce the high background caused by contaminants substantially. Finally, all protein complex purification can be processed under the same conditions, thus the results are reproducible and comparable, which is significant in large-scale systematic proteome researches. Due to these advantages, the TAP method has been successfully applied in the research of protein-protein interactions in prokaryotic and eukaryotic cells.

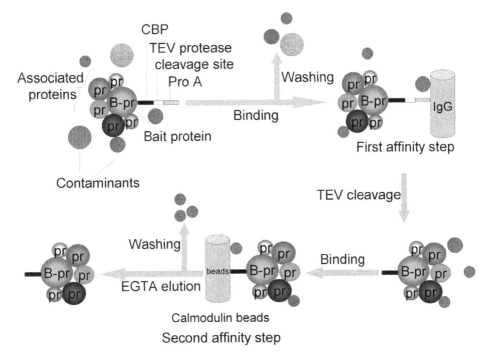

Fig. 2. Schematic of the original TAP method. In the first step, the protein complex, which contains the tagged target protein, combines with an IgG matrix by the ProtA fraction. The protein complex is then eluted using TEV protease under native conditions. In the second step, the elution fraction of the first purification step is incubated with beads coated by calmodulin in the presence of calcium. Subsequently, contaminants and the remainder of TEV protease used in the first step are eliminated through washing. Ultimately, the target protein complex is obtained by elution using EGTA. Adapted from (Xu et al., 2010).

3. The development of TAP tags

Although the TAP system was originally developed in yeast, it has been proven to successfully work in a broad range of organisms. The classic ProtA-TEV-CBP tag may be inefficient to purify all given protein complexes. Therefore several variations of the TAP tag based on other affinity tags have been developed that offer advantages in specific cases (Fig. 1B). The properties of these basic affinity tags (Li, 2010; Lichty et al., 2005; Stevens, 2000; Terpe, 2003) are summarized (Xu et al., 2010) to highlight the advantages and disadvantages of corresponding recombinant tags.

The CBP tag could not always recover protein complexes with high efficient, especially where EGTA may irreversibly interfere with the metal-binding protein function. Consequently, one major type of variation is the replacement of the CBP tag. For example, when purifying protein complexes from mammalian cells growing in monolayer cultures, a biotinylation tag is used as the second affinity tag, taking advantage of the high biotin-avidin binding affinity and resulting in an increased yield of the fusion protein (Drakas et al., 2005). Another example is that the CBP tag has been replaced with a protein C epitope (ProtC) resulting in a new TAP tag, designated PTP (Mani et al., 2011; Schimanski et al., 2005). The advantage of this is that ProtC shows more efficiency and allows the elution either by EGTA or by the ProtC peptide. With respect to isolation of active metal-binding proteins, another replacement of the CBP moiety has been a 9×myc with a 6×His sequence (Rubio et al., 2005). This tag is known as TAPa tag and also contains a human rhinovirus 3C protease cleavage site (HRV 3C) instead of the original TEV site. In contrast to TEV protease, 3C protease still has enzymatic activity at 4℃. These modifications are thought to be beneficial to keep the stabilization of protein complex structures and activities.

Another type of variation is a series TAP tags with smaller size. The original TAP tag is as large as approximately 21 kDa, and this size might impair the function of the tagged protein or interfere with protein complex formation. Because of this, many affinity tags, which range in size from 5–51 amino acids, can be used to replace of CBP or ProtA moiety (Terpe, 2003). One example of a smaller TAP tag is SPA tag, made by substituting 3×FLAG for ProtA (Zeghouf et al., 2004). Replacement of the CBP with a spacer and a single FLAG sequence constitute another smaller tag for TAP (Knuesel et al., 2003). The combination of a streptavidin-binding peptide (SBP) and a CBP has been verified in human cells (Ahmed et al., 2010; Colpitts et al., 2011). Recently, use of another alternative tandem affinity tag, composed of two Strep-tag II and a FLAG-tag (SF), has been published (Gloeckner et al., 2007). This SF tag reduced the size of the TAP tag to 4.6 kDa. This smaller size is less likely to disturb protein activity and structure. Because both tags can be eluted under native conditions, the SF-TAP strategy allows purification of protein complexes in less than 2.5 h. Another similar tandem combination of FLAG-Strep tag II has been developed to purify protein complexes efficiently from *Thiocapsa roseopersicina* (Fodor et al., 2004). And a tandem SBP-FLAG tag has been used to uncover the interacting proteins from HEK293 cells (Zhao et al., 2011). These FLAG-containing combination tags may take advantage of shorter length of the tag and result in higher purity of fusion proteins, while the disadvantage of FLAG tag is the relatively high cost during purification. Lehmann et al. (Lehmann et al., 2009) developed a novel S3S tag comprising a S-tag, a HRV 3C and a Strep-tag II. The S3S tag with a size of 4.2 kDa fulfils the requirements of specificity, high yield and no adverse effects on protein function. Nevertheless, it is doubtful as to whether large tags actually disturb the function of tagged proteins. It would appear that the majority of proteins tagged with the original protA-TEV-CBP tag remain functional, and even small proteins such as acyl-carrier protein (< 10 kDa) (Gully et al., 2003) and thioredoxin (~12 kDa) (Kumar et al., 2004) can be used as bait to purify protein complexes.

In addition to those described above, there are a variety of other TAP tags that are largely different from the classic TAP tag. Bürckstümmer et al. (Bürckstümmer et al., 2006) designed a new TAP tag, designated as GS tag. This tag comprised two copies of IgG binding units of protein G from *Streptococcus* sp. (ProtG) and a SBP. The GS tag was able to purify recombinant proteins with high efficiency and purity, however, the size of the GS tag, at approximately 19 kDa, might be the obvious disadvantage. In a recently published paper, a new tandem affinity tag, the HB tag (Guerrero et al., 2006), consisting of two 6×His motifs and a biotinylation signal peptide has been developed. The HB tag is compatible with *in vivo* cross-linking to purify protein complexes under fully denaturing conditions, which may be beneficial to detect transient and weak protein-protein interactions. A useful derivative of the HB tag is the HTB tag, which includes a TEV cleavage site allowing for protease-driven elution from streptavidin resins (Tagwerker et al., 2006). A CHH tag consisting of a CBP, 6×His residues and three copies of the hemagglutinin (3×HA) has been designed (Honey et al., 2001). However, in fact, the 3×HA peptide is usually used to detect the expression levels of tagged proteins rather than act as the third purification step. In practice, the elution buffer for the calmodulin resin is incompatible with binding to the Ni^{2+} resin. Although buffer exchange may solve this problem, it results in a significant loss of yield. For this reason, the combination of CBP and His tags is generally not recommended. As for purification of associated proteins from *Drosophila* tissues, the 3×FLAG-6×His tag provided significant higher yields than the tranditional tag (Yang et al., 2006). At the same time, a similar combination of His and FLAG epitope was constructed to isolate protein complexes from pathogenic fungus (Kaneko et al., 2004). The HPM tag, another bipartite affinity tag, consisting of 9×His, 9×myc epitope and two copies of HRV 3C inserted was successfully applied in yeast (Graumann et al., 2004).

4. Application of the TAP method

With the development of the TAP approach over the past decade, this method has been employed in the analysis of protein-protein interactions and protein complexes in many different organisms, including yeast, mammals, plants, *Drosophila* and bacteria (Table 1) (Chang, 2006; Xu et al., 2010).

4.1 TAP in yeast

The TAP method was originally developed for analysis of protein complexes in yeast at near-physiological conditions. Gavin et al. (Gavin et al., 2006; Gavin et al., 2002) and Krogan er al. (Krogan et al., 2006) utilized TAP in the large-scale analysis of multi-protein complexes in *Saccharomyces cerevisiae*, in which hundreds to thousands tagged proteins were successfully purified and the associated proteins and involved protein-protein interactions were identified. These results give the possibility to intensive study the functional and organizational network of proteins in yeast.

As to a given protein, the TAP system could also provide opportunity to investigate protein interaction (Graumann et al., 2004; Guerrero et al., 2006; Honey et al., 2001; Krogan et al., 2002). For example, TAP analysis revealed more than one hundred previously known and possible interacting proteins for 21 tagged proteins, which are involved in transcription and progression during mitosis (Graumann et al., 2004). In addition, an active Clb2-Cdc28 kinase complex was purified from yeast cell lysate by TAP (Honey et al., 2001), and four proteins were identified by mass spectrometry to be associated with this complex.

The application of TAP protocol was successful not only in *S. cerevisiae*, but also in *Schizosaccharomyces pombe* and *Candida albicans*. A large number of researches have carry out the TAP strategy to isolate protein complexes and associated partners (Cipak et al., 2009; Gould et al., 2004; Kaneko et al., 2004; Tasto et al., 2001).

Organism	Protein complex	Bait protein	Tandem tag	Proteins identified	Functional pathway
Yeast	RNA polymerase II elongation factors	15 proteins	Protein A, CBP	25	Transcriptional elongation
	Cyclin-dependent kinase complex	Cdc2p	Protein A, CBP	3	Orchestrating cell cycle
	Arp2/3 complex	Arp2p	Protein A, CBP	6	Nucleation
	Swi 5 contaning complex	Swi 5	Protein A, CBP	4	Mating-type switching and homologous recombination regulation
	Spetin complex	CaCDC11	His, FLAG	4	Cytokinesis, virulence
Human cells	DNA-dependent protein kinase	Ku70, Ku80	Protein G, SBP	Many	Multiple
	Human cytomegalovirus protein kinase	UL97	Protein A, Protein C	1	Human cytomegalovirus replication
	ADAP	ADAP	S tag, Strep II	Many	Integrin adhesion regulation
	Many	Spartin	Protein A, CBP	94	Multiple
	Parkin-associated complex	Parkin	streptavidin-binding peptide, CBP	14	Regulation mitochondrial activity
	Many	32 proteins	Protein A, CBP	Many	mRNA formation
	MCM complex	MCM-BP	Protein A, CBP	5	Initiation of DNA replication
	IRF-4 containing complex	IRF-4	Protein A, CBP	1	Activiates IL-2 and IL-4 promoters
	Flavivirus capsid containing complex	Flavivirus capsid	SBP, CBP	4	Disruption of nucleosome formation

Organism	Protein complex	Bait protein	Tandem tag	Proteins identified	Functional pathway
	HDGF	HDGF	SBP, FLAG	106	Multiple
	HP1 containing complex	HP1α, β, γ	Protein A, CBP	Many	Multiple
	Gβγ-Rap1-Radil complex	Gβ2, Gγ2	Strep, CBP	2	Cell-matrix adhesion
Nicotiana benthamiana	Resistance protein N containing complex	Resistance protein N	ProteinA, Myc, His	1	Plant defense
Arabidopsis	Many	6 proteins	Protein A, CBP	42	Cell division signaling
Rice	Many	45 proteins	Protein A, CBP	Many	Multiple
Drosophila	Notch signaling complex	4 proteins	Protein A, CBP	400	Signaling pathway
Escherichia coli	Thioredoxin associated complex	Thioredoxin	Protein A, CBP	80	Multiple
	RecQ complex	RecQ	Protein A, CBP	3	DNA unwinding
	RNP	SrmB	Protein A, CBP	2	Ribosome assembly
Dictyostelium	Microtubule-associated protein	DdEB1, DdCP224	Protein A, CBP	Many	Multiple
	Arp2/3 complex	pArc-34	Protein A, CBP	6	Nucleation
Trypanosome brucei	RNA polymerase I	RPB6z	Protein A, Protein C	3	Transcription
	Alba containing complex	Alba1, 2, 3, 4	Protein A, Protein C	Many	Regulation of translation
Plasmodium falciparum	Translation elongation factor	PfEF-1β	Protein A, Protein C	4	Translation elongation

Table 1. Representative applications of the TAP method.

4.2 TAP in mammalian systems

The application of TAP method has made considerable progress in mammalian systems (Bürckstümmer et al., 2006; Davison et al., 2009; Drakas et al., 2005; Gottlieb and Jackson, 1993; Holowaty et al., 2003; Jeronimo et al., 2007; Kamil and Coen, 2007; Knuesel et al., 2003; Lehmann et al., 2009; Milewska et al., 2009; Sakwe et al., 2007). For instance, human active SMAD3 protein complex was purified from cell lysates through TAP method, and HSP70 was identified as a novel combination partner of SMAD3 (Knuesel et al., 2003). However, in this research the TAP system took a FLAG tag for the second purification step, the elution conditions of which was incompatible with the liquid chromatography-MS/MS sequence application. Therefore, an additional purification step might be reqired to resolve this trouble, which would be time consuming and lead to more sample loss. Sakwe et al. identified a new form of the minichromosome maintenance (MCM) complex in human cells (Sakwe et al., 2007). In another study using the TAP process, Holowaty et al. (Holowaty et al., 2003) expressed Epstein-Barr nuclear antigen-1 (EBNA1) protein in fusion with a TAP tag at the C-terminus in human 293T cells. Several specific cellular protein interactions and some important regulating proteins were discovered. The TAP method could also be used in analysis nuclear protein interaction. The specific association of interferon regulatory factor (IRF)-4 with c-Rel was revealed from human HUT102 cells (Shindo et al., 2011). A directed proteomic analysis of heterochromatin protein 1 (HP1) isotypes interacting partners were identified by the TAP approach (Rosnoblet et al., 2011). Using TAP strategy, more than one hundred proteins were found to interact with hepatoma-derived growth factor (HDGF) in human HEK293 cells (Zhao et al., 2011). The relationship between HDGF and associated proteins suggests that DHGF as a multifunctional molecular might be involved in many cellular activities. In human HEK293 cells, Gβγ subunits of heterotrimeric G proteins were identified to enter in a protein complex with Rap1a and its effector Radil (Ahmed et al., 2010). This result sggested that the Gβγ -Rap1-Radil complex played an important role in cell adhesion.Besides, TAP also allows for purification of protein complexes from mouse fibroblast cells growing in monolayer cultures and mouse embryonic stem cells (Drakas et al., 2005; Mak et al., 2010). It is meaningful that the TAP system could be used to analysis the interaction of virus and host cells during the infection procedure (Colpitts et al., 2011).

4.3 TAP in plants

Recent studies have shown that the TAP strategy is useful in plant protein complex analysis. The first report of the purification of protein complexes from plant tissue by the TAP method was published in 2004 by Rohila et al (Rohila et al., 2004). By using a TAP-tagged hybrid transcription factor as bait, HSP70 and HSP60 were co-purified. This result was verified by former reports (Dittmar et al., 1997; Stancato et al., 1996). Through the TAP strategy, Liu et al. demonstrated that Hsp90 associated with the plant resistance protein N (Liu et al., 2004), which meant that Hsp90 plays an important role in plant defense (Kanzaki et al., 2003; Takahashi et al., 2003). In another study, the Cf-9 protein function in initiating defense signaling was also investigated by TAP (Rivas et al., 2002). The TAP applications described above were all carried out in a transient expression system of *Nicotiana benthamiana*.

On the other hand, the TAP system was utilized to purify a protein complex in stable expression system of *Arabidopsis thaliana*, for the first time in 2005 (Rubio et al., 2005). The components of the target protein complex were all co-purified with the tagged bait. The superiority of this TAP strategy is based on a constitutive promoter, which allowed for over-expression of TAP fusion proteins. The strength of this method is that over-expression increases incorporation of the tagged protein into a protein complex, when the tagged protein is the core component of a complex or a mutant and suppressed expression for the target protein is harmful to cells. In 2006, Brown et al. (Brown et al., 2006) utilized TAP tagged fatty acid synthase components to investigate protein interactions *in vivo* from stably transfected *A. thaliana*. In addition to the application of TAP to a whole, *A. thaliana* cell suspension culture is ideal for investigating protein-protein interactions involved in cell cycle (Van Leene et al., 2007).

Purification of protein complexes by TAP was demonstrated to be effective in rice (Rohila et al., 2006; Rohila et al., 2009), suggesting that the TAP method could be utilized in cereal crops.

4.4 TAP in Drosophila

In 2003 Forler et al. (Forler et al., 2003) successfully expressed TAP-tagged human proteins and purified their *Drosophila melanogaster* (Dm) binding partners in Dm Schneider cells. The critical advantage in this system is the introduction of RNA interference (RNAi), which can suppress the expression of the corresponding endogenous proteins, thereby avoiding competition from them during protein complex assembly. But the complexes purified through this system consisted of two different source proteins, human bait protein and Dm binding partners, therefore the reliability of the interaction needed validation with other experimental strategies. Both in *Drosophila* cultured cells and embryos, several components of the Notch signaling pathway were tagged with a TAP tag and many novel interactions were uncovered (Veraksa et al., 2005). Throughout the TAP progress, Hsc70 and Hsp83 were validated as cofactors of the *Drosophila* nuclear receptor protein for the first time (Yang et al., 2006).

4.5 TAP in bacteria

In recent years, with the development of the TAP procedure, the application of TAP was extended to purification of protein complexes from bacteria. Gully et al. (Gully et al., 2003) first used the TAP protocol in *E. coli* to isolate native protein complexes. Kumar et al. (Kumar et al., 2004) have identified 80 proteins associated with thioredoxin in *E. coli* suggesting multifunction of thioredoxin. Shereda et al. (Shereda et al., 2007) employed the TAP approach to purify the RecQ complexes, and three heterologous proteins were identified. On the basis of the amount of these three binding proteins, these interactions were classed as direct or indirect. This may imply a new application aspect of TAP in interaction identification. SrmB, one of the five *E. coli* DEAD-box proteins was discovered to form a specific ribonucleoprotein (RNP) complex with r-proteins L4, L24 and the 5' region of 23S rRNA using the TAP procedure (Trubetskoy et al., 2009). Similar to the application of the TAP method in global protein complexes analysis in yeast, a large-scale analysis of protein complexes, which revealed a novel protein interaction network in *E. coli*, was reported (Butland et al., 2005). Besides the application of the TAP method in *E. coli*, TAP was also carried out in *Thiocapsa roseopersicina* (Fodor et al., 2004) and *Bacillus subtilis* (Yang et al., 2008).

4.6 TAP in other organisms

The efficiency of the TAP method in purification of protein complexes and identification of interactions was also tested in other organisms, including *Dictyostelium* (Koch et al., 2006; Meima et al., 2007), *Trypanosoma bruce* (Mani et al., 2011; Nguyen et al., 2007; Palfi et al., 2005; Schimanski et al., 2005; Walgraffe et al., 2005) and *Plasmodium falciparum* (Takebe et al., 2007).

5. Problems and future prospects

The TAP method has been successfully used for purification and identification of protein complexes and interacting components both in prokaryotic and eukaryotic organisms. However, in practice, the application effects of the method may have been influenced by its inherentvice. Gavin et al. (Gavin et al., 2002) found that in their large-scale analysis of yeast proteome, not all of the tagged proteins could be purified and not all of the purified tagged proteins could interact with other proteins. They ascribed this failure to the intrinsic quality of the TAP tag. The TAP tag fused to a target protein may interfere with protein function, location and interactions (Mak et al., 2010). One of the possible solutions is to add the tag at the other terminus of the ORFs or to replace the original tag with another one. The CBP affinity step has been proved to be problematic in that case where many endogenous proteins of mammalian cells interact with calmodulin in a calcium-dependent manner (Agell et al., 2002; Head, 1992). A simple alternative solution is replacing the CBP tag with other affinity tags, such as the FLAG sequence (Gloeckner et al., 2007; Knuesel et al., 2003), ProtC (Schimanski et al., 2005) and biotinylation tag (Drakas et al., 2005). The main challenge of the TAP strategy comes from the competition of endogenous proteins with the tagged protein in protein complex assembly. This can be resolved by using RNAi to reduce the endogenous expression level (Forler et al., 2003).In some cases, when the target protein is essential and a mutant of it might be harmful and lethal, the over-expression strategy is a perfect strategy to obtain a protein complex containing the tagged target protein (Ho et al., 2002; Rohila et al., 2006; Rubio et al., 2005). However, bait overexpression possibly cause the formation of nonbiological interactions. In addition, overespression may affect cell viability or cellular activity (e.g. negative regulators of cell metabolism). On this occasion, an inducible promoter is a viable choice, which allows experimental modulation of target protein expression, both in terms of amount and timing.

It is thought that the TAP approach is not a powerful tool to detect transient interactions. Therefore, an *in vivo* cross-linking step is added to freeze both weak and transient interactions taking place in intact cells before lysis (Guerrero et al., 2006; Rohila et al., 2004). The cross-linking method has been widely used in the investigation of protein-DNA and protein-protein interactions (Hall and Struhl, 2002; Jackson, 1999; Kuo and Allis, 1999; Orlando et al., 1997; Otsu et al., 1994; Schmitt-Ulms et al., 2004; Schmitt-Ulms et al., 2001; Vasilescu et al., 2004).

Although the two sequential purification steps of the TAP method largely reduce the background resulting from non-specific protein binding compared to a single purification step, these contaminants cannot be removed completely. Collins et al. (Collins et al., 2007) have compared the results from the two large-scale studies of protein complexes in yeast (Gavin et al., 2006; Krogan et al., 2006) and found the two datasets shared very low degrees of overlap. The major difference between the two datasets was mainly caused by non-specific interactions. The problem of non-specifically interacting proteins can be overcome by comparing several interaction datasets (Ewing et al., 2007), using stable-isotope labelling by amino acids in cell culture (Blagoev et al., 2003; Mann, 2006) or isotope-coded affinity tag (Ranish et al., 2003), thereby completely eliminating false-positive interactions.

The TAP system is considered to be inefficient in identifying interactions occurring only in special physiological states or those which occur for a short period. Whether the TAP tag impairs protein function and complex assembly also remains largely unknown and speculative. These disadvantages may affect its application in such instances.

6. Conclusion

Understanding protein function is a major goal in biology. Although the TAP method has some inherent shortcomings, it is undoubtedly a reasonable system for use in purification of protein complexes and identification of protein-protein interactions. In addition to identifying interactions between proteins, the TAP method could be used to characterize and verify interactions between protein and DNA or between protein and RNA (Hogg and Collins, 2007; Zhao et al., 2011). For protein-DNA/RNA interaction analysis, the use of benzonase must be avoided, and RNase inhibitors should be added to protect RNA intact. At the same time, the TAP method can also be used to analyze the effect of mutants on protein interaction and association, possibly resulting in the discovery of binding sites. Protein purification under near-physiological conditions through the TAP strategy is compatible with functional studies and this advantage allows for mapping of large-scale functional interaction networks. As the procedures and conditions used during the TAP process do not vary greatly among different proteins, the results that are generated by this method should be compiled in a database in order to provide comparable and detailed information on the potential and confirmed functions of proteins, as well as the composition of protein complexes and even the structure and activity of protein complexes.

7. Abbreviations used

TAP, tandem affinity purification; ProtA, IgG-binding units of protein A of *Staphylococcus aureus*; CBP, calmodulin-binding domain; TEV, tobacco etch virus; ProtC, protein C epitope; HRV 3C, human rhinovirus 3C protease cleavage site; SBP, streptavidin-binding peptide; ProtG, IgG binding units of protein G from *Streptococcus* sp.; HA, hemagglutinin; ADAP, adhesion and degranulation promoting adaptor protein; MCM, minichromosome maintenance; EBNA1, Epstein-Barr nuclear antigen-1; IRF, interferon regulatory factor; HP1, heterochromatin protein 1; HDGF, hepatoma-derived growth factor; Dm, *Drosophila melanogaster*; RNAi, RNA interference; RNP, ribonucleoprotein.

8. References

Agell, N., Bachs, O., Rocamora, N.&Villalonga, P. (2002). Modulation of the Ras/Raf/MEK/ERK pathway by Ca(2+), and calmodulin. *Cell. Signal.*, Vol. 14, No. 8, pp. 649-654.

Ahmed, S. M., Daulat, A. M., Meunier, A.&Angers, S. (2010). G protein betagamma subunits regulate cell adhesion through Rap1a and its effector Radil. *J. Biol. Chem.*, Vol. 285, No. 9, pp. 6538-6551.

Alberts, B. (1998). The cell as a collection of protein machines: preparing the next generation of molecular biologists. *Cell*, Vol. 92, No. 3, pp. 291-294.

Bürckstümmer, T., Bennett, K. L., Preradovic, A., Schutze, G., Hantschel, O., Superti-Furga, G.&Bauch, A. (2006). An efficient tandem affinity purification procedure for interaction proteomics in mammalian cells. *Nat. Methods*, Vol. 3, No. 12, pp. 1013-1019.

Blagoev, B., Kratchmarova, I., Ong, S. E., Nielsen, M., Foster, L. J.&Mann, M. (2003). A proteomics strategy to elucidate functional protein-protein interactions applied to EGF signaling. *Nat. Biotechnol.*, Vol. 21, No. 3, pp. 315-318.

Brown, A. P., Affleck, V., Fawcett, T.&Slabas, A. R. (2006). Tandem affinity purification tagging of fatty acid biosynthetic enzymes in Synechocystis sp. PCC6803 and Arabidopsis thaliana. *J. Exp. Bot.*, Vol. 57, No. 7, pp. 1563-1571.

Butland, G., Peregrin-Alvarez, J. M., Li, J., Yang, W., Yang, X., Canadien, V., Starostine, A., Richards, D., Beattie, B., Krogan, N., Davey, M., Parkinson, J., Greenblatt, J.&Emili, A. (2005). Interaction network containing conserved and essential protein complexes in Escherichia coli. *Nature*, Vol. 433, No. 7025, pp. 531-537.

Chang, I. F. (2006). Mass spectrometry-based proteomic analysis of the epitope-tag affinity purified protein complexes in eukaryotes. *Proteomics*, Vol. 6, No. 23, pp. 6158-6166.

Cipak, L., Spirek, M., Novatchkova, M., Chen, Z., Rumpf, C., Lugmayr, W., Mechtler, K., Ammerer, G., Csaszar, E.&Gregan, J. (2009). An improved strategy for tandem affinity purification-tagging of Schizosaccharomyces pombe genes. *Proteomics*, Vol. 9, No. 20, pp. 4825-4828.

Collins, S. R., Kemmeren, P., Zhao, X. C., Greenblatt, J. F., Spencer, F., Holstege, F. C., Weissman, J. S.&Krogan, N. J. (2007). Toward a comprehensive atlas of the physical interactome of Saccharomyces cerevisiae. *Mol. Cell. Proteomics*, Vol. 6, No. 3, pp. 439-450.

Colpitts, T. M., Barthel, S., Wang, P.&Fikrig, E. (2011). Dengue virus capsid protein binds core histones and inhibits nucleosome formation in human liver cells. *PLoS One*, Vol. 6, No. 9, pp. e24365.

Cusick, M. E., Klitgord, N., Vidal, M.&Hill, D. E. (2005). Interactome: gateway into systems biology. *Hum. Mol. Genet.*, Vol. 14 No. pp. 171-181.

Davison, E. J., Pennington, K., Hung, C. C., Peng, J., Rafiq, R., Ostareck-Lederer, A., Ostareck, D. H., Ardley, H. C., Banks, R. E.&Robinson, P. A. (2009). Proteomic analysis of increased Parkin expression and its interactants provides evidence for a role in modulation of mitochondrial function. *Proteomics*, Vol. 9, No. 18, pp. 4284-4297.

Dittmar, K. D., Demady, D. R., Stancato, L. F., Krishna, P.&Pratt, W. B. (1997). Folding of the glucocorticoid receptor by the heat shock protein (hsp) 90-based chaperone machinery. The role of p23 is to stabilize receptor.hsp90 heterocomplexes formed by hsp90.p60.hsp70. *J. Biol. Chem.*, Vol. 272, No. 34, pp. 21213-21220.

Drakas, R., Prisco, M.&Baserga, R. (2005). A modified tandem affinity purification tag technique for the purification of protein complexes in mammalian cells. *Proteomics*, Vol. 5, No. 1, pp. 132-137.

Ewing, R. M., Chu, P., Elisma, F., Li, H., Taylor, P., Climie, S., McBroom-Cerajewski, L., Robinson, M. D., O'Connor, L., Li, M., Taylor, R., Dharsee, M., Ho, Y., Heilbut, A., Moore, L., Zhang, S., Ornatsky, O., Bukhman, Y. V., Ethier, M., Sheng, Y., Vasilescu, J., Abu-Farha, M., Lambert, J. P., Duewel, H. S., Stewart, II, Kuehl, B., Hogue, K., Colwill, K., Gladwish, K., Muskat, B., Kinach, R., Adams, S. L., Moran, M. F., Morin, G. B., Topaloglou, T.&Figeys, D. (2007). Large-scale mapping of human protein-protein interactions by mass spectrometry. *Mol. Syst. Biol.*, Vol. 3, No. pp. 89-105.

Fodor, B. D., Kovacs, A. T., Csaki, R., Hunyadi-Gulyas, E., Klement, E., Maroti, G., Meszaros, L. S., Medzihradszky, K. F., Rakhely, G.&Kovacs, K. L. (2004). Modular broad-host-range expression vectors for single-protein and protein complex purification. *Appl. Environ. Microbiol.*, Vol. 70, No. 2, pp. 712-721.

Forler, D., Kocher, T., Rode, M., Gentzel, M., Izaurralde, E.&Wilm, M. (2003). An efficient protein complex purification method for functional proteomics in higher eukaryotes. *Nat. Biotechnol.*, Vol. 21, No. 1, pp. 89-92.

Fromont-Racine, M., Rain, J. C.&Legrain, P. (1997). Toward a functional analysis of the yeast genome through exhaustive two-hybrid screens. *Nat. Genet.*, Vol. 16, No. 3, pp. 277-282.

Gavin, A. C., Aloy, P., Grandi, P., Krause, R., Boesche, M., Marzioch, M., Rau, C., Jensen, L. J., Bastuck, S., Dumpelfeld, B., Edelmann, A., Heurtier, M. A., Hoffman, V., Hoefert, C., Klein, K., Hudak, M., Michon, A. M., Schelder, M., Schirle, M., Remor, M., Rudi, T., Hooper, S., Bauer, A., Bouwmeester, T., Casari, G., Drewes, G., Neubauer, G., Rick, J. M., Kuster, B., Bork, P., Russell, R. B.&Superti-Furga, G. (2006). Proteome survey reveals modularity of the yeast cell machinery. *Nature*, Vol. 440, No. 7084, pp. 631-636.

Gavin, A. C., Bosche, M., Krause, R., Grandi, P., Marzioch, M., Bauer, A., Schultz, J., Rick, J. M., Michon, A. M., Cruciat, C. M., Remor, M., Hofert, C., Schelder, M., Brajenovic, M., Ruffner, H., Merino, A., Klein, K., Hudak, M., Dickson, D., Rudi, T., Gnau, V., Bauch, A., Bastuck, S., Huhse, B., Leutwein, C., Heurtier, M. A., Copley, R. R., Edelmann, A., Querfurth, E., Rybin, V., Drewes, G., Raida, M., Bouwmeester, T., Bork, P., Seraphin, B., Kuster, B., Neubauer, G.&Superti-Furga, G. (2002). Functional organization of the yeast proteome by systematic analysis of protein complexes. *Nature*, Vol. 415, No. 6868, pp. 141-147.

Gloeckner, C. J., Boldt, K., Schumacher, A., Roepman, R.&Ueffing, M. (2007). A novel tandem affinity purification strategy for the efficient isolation and characterisation of native protein complexes. *Proteomics*, Vol. 7, No. 23, pp. 4228-4234.

Gottlieb, T. M.&Jackson, S. P. (1993). The DNA-dependent protein kinase: requirement for DNA ends and association with Ku antigen. *Cell*, Vol. 72, No. 1, pp. 131-142.

Gould, K. L., Ren, L., Feoktistova, A. S., Jennings, J. L.&Link, A. J. (2004). Tandem affinity purification and identification of protein complex components. *Methods*, Vol. 33, No. 3, pp. 239-244.

Graumann, J., Dunipace, L. A., Seol, J. H., McDonald, W. H., Yates, J. R., 3rd, Wold, B. J.&Deshaies, R. J. (2004). Applicability of tandem affinity purification MudPIT to pathway proteomics in yeast. *Mol. Cell. Proteomics*, Vol. 3, No. 3, pp. 226-237.

Guerrero, C., Tagwerker, C., Kaiser, P.&Huang, L. (2006). An integrated mass spectrometry-based proteomic approach: quantitative analysis of tandem affinity-purified in vivo cross-linked protein complexes (QTAX) to decipher the 26 S proteasome-interacting network. *Mol. Cell. Proteomics*, Vol. 5, No. 2, pp. 366-378.

Gully, D., Moinier, D., Loiseau, L.&Bouveret, E. (2003). New partners of acyl carrier protein detected in Escherichia coli by tandem affinity purification. *FEBS Lett.*, Vol. 548, No. 1-3, pp. 90-96.

Hall, D. B.&Struhl, K. (2002). The VP16 activation domain interacts with multiple transcriptional components as determined by protein-protein cross-linking in vivo. *J. Biol. Chem.*, Vol. 277, No. 48, pp. 46043-46050.

Head, J. F. (1992). A better grip on calmodulin. *Curr. Biol.*, Vol. 2, No. 11, pp. 609-611.

Ho, Y., Gruhler, A., Heilbut, A., Bader, G. D., Moore, L., Adams, S. L., Millar, A., Taylor, P., Bennett, K., Boutilier, K., Yang, L., Wolting, C., Donaldson, I., Schandorff, S., Shewnarane, J., Vo, M., Taggart, J., Goudreault, M., Muskat, B., Alfarano, C., Dewar, D., Lin, Z., Michalickova, K., Willems, A. R., Sassi, H., Nielsen, P. A., Rasmussen, K. J., Andersen, J. R., Johansen, L. E., Hansen, L. H., Jespersen, H., Podtelejnikov, A., Nielsen, E., Crawford, J., Poulsen, V., Sorensen, B. D.,

Matthiesen, J., Hendrickson, R. C., Gleeson, F., Pawson, T., Moran, M. F., Durocher, D., Mann, M., Hogue, C. W., Figeys, D.&Tyers, M. (2002). Systematic identification of protein complexes in Saccharomyces cerevisiae by mass spectrometry. *Nature*, Vol. 415, No. 6868, pp. 180-183.

Hogg, J. R.&Collins, K. (2007). RNA-based affinity purification reveals 7SK RNPs with distinct composition and regulation. *RNA*, Vol. 13, No. 6, pp. 868-880.

Holowaty, M. N., Zeghouf, M., Wu, H., Tellam, J., Athanasopoulos, V., Greenblatt, J.&Frappier, L. (2003). Protein profiling with Epstein-Barr nuclear antigen-1 reveals an interaction with the herpesvirus-associated ubiquitin-specific protease HAUSP/USP7. *J. Biol. Chem.*, Vol. 278, No. 32, pp. 29987-29994.

Honey, S., Schneider, B. L., Schieltz, D. M., Yates, J. R.&Futcher, B. (2001). A novel multiple affinity purification tag and its use in identification of proteins associated with a cyclin-CDK complex. *Nucleic Acids Res.*, Vol. 29, No. 4, pp. 24-32.

Ito, T., Chiba, T., Ozawa, R., Yoshida, M., Hattori, M.&Sakaki, Y. (2001). A comprehensive two-hybrid analysis to explore the yeast protein interactome. *Proc. Natl. Acad. Sci. USA*, Vol. 98, No. 8, pp. 4569-4574.

Jackson, V. (1999). Formaldehyde cross-linking for studying nucleosomal dynamics. *Methods*, Vol. 17, No. 2, pp. 125-139.

Jeronimo, C., Forget, D., Bouchard, A., Li, Q., Chua, G., Poitras, C., Therien, C., Bergeron, D., Bourassa, S., Greenblatt, J., Chabot, B., Poirier, G. G., Hughes, T. R., Blanchette, M., Price, D. H.&Coulombe, B. (2007). Systematic analysis of the protein interaction network for the human transcription machinery reveals the identity of the 7SK capping enzyme. *Mol. Cell*, Vol. 27, No. 2, pp. 262-274.

Kamil, J. P.&Coen, D. M. (2007). Human cytomegalovirus protein kinase UL97 forms a complex with the tegument phosphoprotein pp65. *J. Virol.*, Vol. 81, No. 19, pp. 10659-10668.

Kaneko, A., Umeyama, T., Hanaoka, N., Monk, B. C., Uehara, Y.&Niimi, M. (2004). Tandem affinity purification of the Candida albicans septin protein complex. *Yeast*, Vol. 21, No. 12, pp. 1025-1033.

Kanzaki, H., Saitoh, H., Ito, A., Fujisawa, S., Kamoun, S., Katou, S., Yoshioka, H.&Terauchi, R. (2003). Cytosolic HSP90 and HSP70 are essential components of INF1-mediated hypersensitive response and non-host resistance to Pseudomonas cichorii in Nicotiana benthamiana. *Mol Plant Pathol*, Vol. 4, No. 5, pp. 383-391.

Knuesel, M., Wan, Y., Xiao, Z., Holinger, E., Lowe, N., Wang, W.&Liu, X. (2003). Identification of novel protein-protein interactions using a versatile mammalian tandem affinity purification expression system. *Mol. Cell. Proteomics*, Vol. 2, No. 11, pp. 1225-1233.

Koch, K. V., Reinders, Y., Ho, T. H., Sickmann, A.&Graf, R. (2006). Identification and isolation of Dictyostelium microtubule-associated protein interactors by tandem affinity purification. *Eur. J. Cell Biol.*, Vol. 85, No. 9-10, pp. 1079-1090.

Krogan, N. J., Cagney, G., Yu, H., Zhong, G., Guo, X., Ignatchenko, A., Li, J., Pu, S., Datta, N., Tikuisis, A. P., Punna, T., Peregrin-Alvarez, J. M., Shales, M., Zhang, X., Davey, M., Robinson, M. D., Paccanaro, A., Bray, J. E., Sheung, A., Beattie, B., Richards, D. P., Canadien, V., Lalev, A., Mena, F., Wong, P., Starostine, A., Canete, M. M., Vlasblom, J., Wu, S., Orsi, C., Collins, S. R., Chandran, S., Haw, R., Rilstone, J. J., Gandi, K., Thompson, N. J., Musso, G., St Onge, P., Ghanny, S., Lam, M. H.,

Butland, G., Altaf-Ul, A. M., Kanaya, S., Shilatifard, A., O'Shea, E., Weissman, J. S., Ingles, C. J., Hughes, T. R., Parkinson, J., Gerstein, M., Wodak, S. J., Emili, A.&Greenblatt, J. F. (2006). Global landscape of protein complexes in the yeast Saccharomyces cerevisiae. *Nature*, Vol. 440, No. 7084, pp. 637-643.

Krogan, N. J., Kim, M., Ahn, S. H., Zhong, G., Kobor, M. S., Cagney, G., Emili, A., Shilatifard, A., Buratowski, S.&Greenblatt, J. F. (2002). RNA polymerase II elongation factors of Saccharomyces cerevisiae: a targeted proteomics approach. *Mol. Cell. Biol.*, Vol. 22, No. 20, pp. 6979-6992.

Kumar, J. K., Tabor, S.&Richardson, C. C. (2004). Proteomic analysis of thioredoxin-targeted proteins in Escherichia coli. *Proc. Natl. Acad. Sci. USA*, Vol. 101, No. 11, pp. 3759-3764.

Kuo, M. H.&Allis, C. D. (1999). In vivo cross-linking and immunoprecipitation for studying dynamic Protein:DNA associations in a chromatin environment. *Methods*, Vol. 19, No. 3, pp. 425-433.

Lehmann, R., Meyer, J., Schuemann, M., Krause, E.&Freund, C. (2009). A novel S3S-TAP-tag for the isolation of T-cell interaction partners of adhesion and degranulation promoting adaptor protein. *Proteomics*, Vol. 9, No. 23, pp. 5288-5295.

Li, Y. F. (2010). commonly used tag combinations for tandem affinity purification. *Biotechnol. Appl. Biochem.*, Vol. 55, No. pp. 73-83.

Lichty, J. J., Malecki, J. L., Agnew, H. D., Michelson-Horowitz, D. J.&Tan, S. (2005). Comparison of affinity tags for protein purification. *Protein Expr. Purif.*, Vol. 41, No. 1, pp. 98-105.

Liu, Y., Burch-Smith, T., Schiff, M., Feng, S.&Dinesh-Kumar, S. P. (2004). Molecular chaperone Hsp90 associates with resistance protein N and its signaling proteins SGT1 and Rar1 to modulate an innate immune response in plants. *J. Biol. Chem.*, Vol. 279, No. 3, pp. 2101-2108.

Mak, A. B., Ni, Z., Hewel, J. A., Chen, G. I., Zhong, G., Karamboulas, K., Blakely, K., Smiley, S., Marcon, E., Roudeva, D., Li, J., Olsen, J. B., Wan, C., Punna, T., Isserlin, R., Chetyrkin, S., Gingras, A. C., Emili, A., Greenblatt, J.&Moffat, J. (2010). A lentiviral functional proteomics approach identifies chromatin remodeling complexes important for the induction of pluripotency. *Mol. Cell. Proteomics*, Vol. 9, No. 5, pp. 811-823.

Mani, J., Guttinger, A., Schimanski, B., Heller, M., Acosta-Serrano, A., Pescher, P., Spath, G.&Roditi, I. (2011). Alba-domain proteins of Trypanosoma brucei are cytoplasmic RNA-binding proteins that interact with the translation machinery. *PLoS One*, Vol. 6, No. 7, pp. e22463.

Mann, M. (2006). Functional and quantitative proteomics using SILAC. *Nat. Rev. Mol. Cell Biol.*, Vol. 7, No. 12, pp. 952-958.

Meima, M. E., Weening, K. E.&Schaap, P. (2007). Vectors for expression of proteins with single or combinatorial fluorescent protein and tandem affinity purification tags in Dictyostelium. *Protein Expr. Purif.*, Vol. 53, No. 2, pp. 283-288.

Milewska, M., McRedmond, J.&Byrne, P. C. (2009). Identification of novel spartin-interactors shows spartin is a multifunctional protein. *J. Neurochem.*, Vol. 111, No. 4, pp. 1022-1030.

Nguyen, T. N., Schimanski, B.&Gunzl, A. (2007). Active RNA polymerase I of Trypanosoma brucei harbors a novel subunit essential for transcription. *Mol. Cell Biol.*, Vol. 27, No. 17, pp. 6254-6263.

Orlando, V., Strutt, H.&Paro, R. (1997). Analysis of chromatin structure by in vivo formaldehyde cross-linking. *Methods*, Vol. 11, No. 2, pp. 205-214.

Otsu, M., Omura, F., Yoshimori, T.&Kikuchi, M. (1994). Protein disulfide isomerase associates with misfolded human lysozyme in vivo. *J. Biol. Chem.*, Vol. 269, No. 9, pp. 6874-6877.

Palfi, Z., Schimanski, B., Gunzl, A., Lucke, S.&Bindereif, A. (2005). U1 small nuclear RNP from Trypanosoma brucei: a minimal U1 snRNA with unusual protein components. *Nucleic Acids Res.*, Vol. 33, No. 8, pp. 2493-2503.

Puig, O., Caspary, F., Rigaut, G., Rutz, B., Bouveret, E., Bragado-Nilsson, E., Wilm, M.&Seraphin, B. (2001). The tandem affinity purification (TAP) method: a general procedure of protein complex purification. *Methods*, Vol. 24, No. 3, pp. 218-229.

Ranish, J. A., Yi, E. C., Leslie, D. M., Purvine, S. O., Goodlett, D. R., Eng, J.&Aebersold, R. (2003). The study of macromolecular complexes by quantitative proteomics. *Nat. Genet.*, Vol. 33, No. 3, pp. 349-355.

Rigaut, G., Shevchenko, A., Rutz, B., Wilm, M., Mann, M.&Seraphin, B. (1999). A generic protein purification method for protein complex characterization and proteome exploration. *Nat. Biotechnol.*, Vol. 17, No. 10, pp. 1030-1032.

Rivas, S., Romeis, T.&Jones, J. D. (2002). The Cf-9 disease resistance protein is present in an approximately 420-kilodalton heteromultimeric membrane-associated complex at one molecule per complex. *Plant Cell*, Vol. 14, No. 3, pp. 689-702.

Rohila, J. S., Chen, M., Cerny, R.&Fromm, M. E. (2004). Improved tandem affinity purification tag and methods for isolation of protein heterocomplexes from plants. *Plant J.*, Vol. 38, No. 1, pp. 172-181.

Rohila, J. S., Chen, M., Chen, S., Chen, J., Cerny, R., Dardick, C., Canlas, P., Xu, X., Gribskov, M., Kanrar, S., Zhu, J. K., Ronald, P.&Fromm, M. E. (2006). Protein-protein interactions of tandem affinity purification-tagged protein kinases in rice. *Plant J.*, Vol. 46, No. 1, pp. 1-13.

Rohila, J. S., Chen, M., Chen, S., Chen, J., Cerny, R. L., Dardick, C., Canlas, P., Fujii, H., Gribskov, M., Kanrar, S., Knoflicek, L., Stevenson, B., Xie, M., Xu, X., Zheng, X., Zhu, J. K., Ronald, P.&Fromm, M. E. (2009). Protein-protein interactions of tandem affinity purified protein kinases from rice. *PLoS One*, Vol. 4, No. 8, pp. e6685.

Rosnoblet, C., Vandamme, J., Volkel, P.&Angrand, P. O. (2011). Analysis of the human HP1 interactome reveals novel binding partners. *Biochem. Biophys. Res. Commun.*, Vol. 413, No. 2, pp. 206-211.

Rubio, V., Shen, Y., Saijo, Y., Liu, Y., Gusmaroli, G., Dinesh-Kumar, S. P.&Deng, X. W. (2005). An alternative tandem affinity purification strategy applied to *Arabidopsis* protein complex isolation. *Plant J.*, Vol. 41, No. 5, pp. 767-778.

Sakwe, A. M., Nguyen, T., Athanasopoulos, V., Shire, K.&Frappier, L. (2007). Identification and characterization of a novel component of the human minichromosome maintenance complex. *Mol. Cell Biol.*, Vol. 27, No. 8, pp. 3044-3055.

Schimanski, B., Nguyen, T. N.&Gunzl, A. (2005). Characterization of a multisubunit transcription factor complex essential for spliced-leader RNA gene transcription in Trypanosoma brucei. *Mol. Cell. Biol.*, Vol. 25, No. 16, pp. 7303-7313.

Schimanski, B., Nguyen, T. N.&Gunzl, A. (2005). Highly efficient tandem affinity purification of trypanosome protein complexes based on a novel epitope combination. *Eukaryot Cell*, Vol. 4, No. 11, pp. 1942-1950.

Schmitt-Ulms, G., Hansen, K., Liu, J., Cowdrey, C., Yang, J., DeArmond, S. J., Cohen, F. E., Prusiner, S. B.&Baldwin, M. A. (2004). Time-controlled transcardiac perfusion cross-linking for the study of protein interactions in complex tissues. *Nat. Biotechnol.*, Vol. 22, No. 6, pp. 724-731.

Schmitt-Ulms, G., Legname, G., Baldwin, M. A., Ball, H. L., Bradon, N., Bosque, P. J., Crossin, K. L., Edelman, G. M., DeArmond, S. J., Cohen, F. E.&Prusiner, S. B. (2001). Binding of neural cell adhesion molecules (N-CAMs) to the cellular prion protein. *J. Mol. Biol.*, Vol. 314, No. 5, pp. 1209-1225.

Shereda, R. D., Bernstein, D. A.&Keck, J. L. (2007). A central role for SSB in Escherichia coli RecQ DNA helicase function. *J. Biol. Chem.*, Vol. 282, No. 26, pp. 19247-19258.

Shindo, H., Yasui, K., Yamamoto, K., Honma, K., Yui, K., Kohno, T., Ma, Y., Chua, K. J., Kubo, Y., Aihara, H., Ito, T., Nagayasu, T., Matsuyama, T.&Hayashi, H. (2011). Interferon regulatory factor-4 activates IL-2 and IL-4 promoters in cooperation with c-Rel. *Cytokine*, Vol. No. pp.

Stancato, L. F., Hutchison, K. A., Krishna, P.&Pratt, W. B. (1996). Animal and plant cell lysates share a conserved chaperone system that assembles the glucocorticoid receptor into a functional heterocomplex with hsp90. *Biochemistry*, Vol. 35, No. 2, pp. 554-561.

Stevens, R. C. (2000). Design of high-throughput methods of protein production for structural biology. *Structure*, Vol. 8, No. 9, pp. 177-185.

Tagwerker, C., Zhang, H., Wang, X., Larsen, L. S., Lathrop, R. H., Hatfield, G. W., Auer, B., Huang, L.&Kaiser, P. (2006). HB tag modules for PCR-based gene tagging and tandem affinity purification in Saccharomyces cerevisiae. *Yeast*, Vol. 23, No. 8, pp. 623-632.

Takahashi, A., Casais, C., Ichimura, K.&Shirasu, K. (2003). HSP90 interacts with RAR1 and SGT1 and is essential for RPS2-mediated disease resistance in Arabidopsis. *Proc. Natl. Acad. Sci. USA*, Vol. 100, No. 20, pp. 11777-11782.

Takebe, S., Witola, W. H., Schimanski, B., Gunzl, A.&Ben Mamoun, C. (2007). Purification of components of the translation elongation factor complex of Plasmodium falciparum by tandem affinity purification. *Eukaryot Cell*, Vol. 6, No. 4, pp. 584-591.

Tasto, J. J., Carnahan, R. H., McDonald, W. H.&Gould, K. L. (2001). Vectors and gene targeting modules for tandem affinity purification in Schizosaccharomyces pombe. *Yeast*, Vol. 18, No. 7, pp. 657-662.

Terpe, K. (2003). Overview of tag protein fusions: from molecular and biochemical fundamentals to commercial systems. *Appl. Microbiol. Biotechnol.*, Vol. 60, No. 5, pp. 523-533.

Trubetskoy, D., Proux, F., Allemand, F., Dreyfus, M.&Iost, I. (2009). SrmB, a DEAD-box helicase involved in Escherichia coli ribosome assembly, is specifically targeted to 23S rRNA in vivo. *Nucleic Acids Res.*, Vol. 37, No. 19, pp. 6540-6549.

Uetz, P., Giot, L., Cagney, G., Mansfield, T. A., Judson, R. S., Knight, J. R., Lockshon, D., Narayan, V., Srinivasan, M., Pochart, P., Qureshi-Emili, A., Li, Y., Godwin, B., Conover, D., Kalbfleisch, T., Vijayadamodar, G., Yang, M., Johnston, M., Fields,

S.&Rothberg, J. M. (2000). A comprehensive analysis of protein-protein interactions in Saccharomyces cerevisiae. *Nature*, Vol. 403, No. 6770, pp. 623-627.

Van Leene, J., Stals, H., Eeckhout, D., Persiau, G., Van De Slijke, E., Van Isterdael, G., De Clercq, A., Bonnet, E., Laukens, K., Remmerie, N., Henderickx, K., De Vijlder, T., Abdelkrim, A., Pharazyn, A., Van Onckelen, H., Inze, D., Witters, E.&De Jaeger, G. (2007). A tandem affinity purification-based technology platform to study the cell cycle interactome in Arabidopsis thaliana. *Mol. Cell. Proteomics*, Vol. 6, No. 7, pp. 1226-1238.

Vasilescu, J., Guo, X.&Kast, J. (2004). Identification of protein-protein interactions using in vivo cross-linking and mass spectrometry. *Proteomics*, Vol. 4, No. 12, pp. 3845-3854.

Veraksa, A., Bauer, A.&Artavanis-Tsakonas, S. (2005). Analyzing protein complexes in Drosophila with tandem affinity purification-mass spectrometry. *Dev. Dyn.*, Vol. 232, No. 3, pp. 827-834.

Walgraffe, D., Devaux, S., Lecordier, L., Dierick, J. F., Dieu, M., Van den Abbeele, J., Pays, E.&Vanhamme, L. (2005). Characterization of subunits of the RNA polymerase I complex in Trypanosoma brucei. *Mol. Biochem. Parasitol.*, Vol. 139, No. 2, pp. 249-260.

Xu, X., Song, Y., Li, Y., Chang, J., Zhang, H.&An, L. (2010). The tandem affinity purification method: an efficient system for protein complex purification and protein interaction identification. *Protein Expr. Purif.*, Vol. 72, No. 2, pp. 149-156.

Yang, P., Sampson, H. M.&Krause, H. M. (2006). A modified tandem affinity purification strategy identifies cofactors of the Drosophila nuclear receptor dHNF4. *Proteomics*, Vol. 6, No. 3, pp. 927-935.

Yang, X., Doherty, G. P.&Lewis, P. J. (2008). Tandem affinity purification vectors for use in gram positive bacteria. *Plasmid*, Vol. 59, No. 1, pp. 54-62.

Zeghouf, M., Li, J., Butland, G., Borkowska, A., Canadien, V., Richards, D., Beattie, B., Emili, A.&Greenblatt, J. F. (2004). Sequential Peptide Affinity (SPA) system for the identification of mammalian and bacterial protein complexes. *J. Proteome Res.*, Vol. 3, No. 3, pp. 463-468.

Zhao, J., Yu, H., Lin, L., Tu, J., Cai, L., Chen, Y., Zhong, F., Lin, C., He, F.&Yang, P. (2011). Interactome study suggests multiple cellular functions of hepatoma-derived growth factor (HDGF). *J Proteomics*, Vol. No. pp.

Zhu, H., Bilgin, M., Bangham, R., Hall, D., Casamayor, A., Bertone, P., Lan, N., Jansen, R., Bidlingmaier, S., Houfek, T., Mitchell, T., Miller, P., Dean, R. A., Gerstein, M.&Snyder, M. (2001). Global analysis of protein activities using proteome chips. *Science*, Vol. 293, No. 5537, pp. 2101-2105.

Permissions

The contributors of this book come from diverse backgrounds, making this book a truly international effort. This book will bring forth new frontiers with its revolutionizing research information and detailed analysis of the nascent developments around the world.

We would like to thank Jianfeng Cai, for lending his expertise to make the book truly unique. He has played a crucial role in the development of this book. Without his invaluable contribution this book wouldn't have been possible. He has made vital efforts to compile up to date information on the varied aspects of this subject to make this book a valuable addition to the collection of many professionals and students.

This book was conceptualized with the vision of imparting up-to-date information and advanced data in this field. To ensure the same, a matchless editorial board was set up. Every individual on the board went through rigorous rounds of assessment to prove their worth. After which they invested a large part of their time researching and compiling the most relevant data for our readers. Conferences and sessions were held from time to time between the editorial board and the contributing authors to present the data in the most comprehensible form. The editorial team has worked tirelessly to provide valuable and valid information to help people across the globe.

Every chapter published in this book has been scrutinized by our experts. Their significance has been extensively debated. The topics covered herein carry significant findings which will fuel the growth of the discipline. They may even be implemented as practical applications or may be referred to as a beginning point for another development. Chapters in this book were first published by InTech; hereby published with permission under the Creative Commons Attribution License or equivalent.

The editorial board has been involved in producing this book since its inception. They have spent rigorous hours researching and exploring the diverse topics which have resulted in the successful publishing of this book. They have passed on their knowledge of decades through this book. To expedite this challenging task, the publisher supported the team at every step. A small team of assistant editors was also appointed to further simplify the editing procedure and attain best results for the readers.

Our editorial team has been hand-picked from every corner of the world. Their multi-ethnicity adds dynamic inputs to the discussions which result in innovative outcomes. These outcomes are then further discussed with the researchers and contributors who give their valuable feedback and opinion regarding the same. The feedback is then collaborated with the researches and they are edited in a comprehensive manner to aid the understanding of the subject.

Apart from the editorial board, the designing team has also invested a significant amount of their time in understanding the subject and creating the most relevant covers. They scrutinized every image to scout for the most suitable representation of the subject and create an appropriate cover for the book.

The publishing team has been involved in this book since its early stages. They were actively engaged in every process, be it collecting the data, connecting with the contributors or procuring relevant information. The team has been an ardent support to the editorial, designing and production team. Their endless efforts to recruit the best for this project, has resulted in the accomplishment of this book. They are a veteran in the field of academics and their pool of knowledge is as vast as their experience in printing. Their expertise and guidance has proved useful at every step. Their uncompromising quality standards have made this book an exceptional effort. Their encouragement from time to time has been an inspiration for everyone.

The publisher and the editorial board hope that this book will prove to be a valuable piece of knowledge for researchers, students, practitioners and scholars across the globe.

List of Contributors

Catherine H. Kaschula and M. Iqbal Parker
Department of Medical Biochemistry, University of Cape Town, Anzio Road, Observatory, Cape Town, South Africa
International Centre for Genetic Engineering and Biotechnology, Wernher and Beit South, Anzio Rd, Observatory, Cape Town, South Africa

Dirk Lang
Department of Human Biology, University of Cape Town, Anzio Road, Observatory, Cape Town, South Africa

Shun-Ichiro Iemura and Tohru Natsume
Biomedicinal Information Research Center (BIRC), National Institute of Advanced Industrial Science and Technology (AIST), Japan

Sabine Hunke and Volker S. Müller
Molekulare Mikrobiologie, Universität Osnabrück, Osnabrück, Germany

Marco Ambriz-Rivas and Gabriel del Rio
Universidad Nacional Autónoma de México, Instituto de Fisiología Celular, México

Nina Pastor
Universidad Autónoma del Estado de Morelos, Facultad de Ciencias, México

Esther Zúñiga-Sánchez and Alicia Gamboa-de Buen
Universidad Nacional Autónoma de México, México

Youhong Niu, Yaogang Hu, Rongsheng E. Wang, Haifan Wu and Jianfeng Cai
Department of Chemistry, University of South Florida, Tampa, FL, USA

Xiaolong Li and Jiandong Chen
Department of Molecular Oncology, H. Lee Moffitt Cancer Center & Research Institute, Tampa, FL, USA

Daniel Some and Sophia Kenrick
Wyatt Technology Corp., USA

Guy Nir, Moshe Lindner and Yuval Garini
Physics Department and institute of Nanotechnology, Bar Ilan University, Ramat Gan, Israel

Johann P. Klare
Physics Department, University of Osnabrück, Osnabrück, Germany

Xiaoli Xu
Department of Burn and Plastic Surgery, Tangdu Hospital, Fourth Military Medical University, Xi'an, China
Key Laboratory of Arid and Grassland Agroecology of Ministry of Education, School of Life Sciences, Lanzhou University, Lanzhou, China

Xueyong Li
Department of Burn and Plastic Surgery, Tangdu Hospital, Fourth Military Medical University, Xi'an, China

Hua Zhang and Lizhe An
Key Laboratory of Arid and Grassland Agroecology of Ministry of Education, School of Life Sciences, Lanzhou University, Lanzhou, China

Printed in the USA
CPSIA information can be obtained
at www.ICGtesting.com
JSHW011415221024
72173JS00004B/550